People-Centric Project Management

by Richard C. Bernheim, PMP, MBA

First Edition

Oshawa, Ontario

People-Centric Project Management
by Richard C. Bernheim, PMP, MBA

Managing Editor: Kevin Aguanno
Typesetting: Charles Sin
Cover Design: Troy O'Brien
eBook Conversion: Charles Sin and Agustina Baid

Published by:

Multi-Media Publications Inc.
Box 58043, Rosslynn RPO
Oshawa, ON, Canada, L1J 8L6

http://www.mmpubs.com/

Paperback   ISBN-13: 9781554891047
Adobe PDF ebook   ISBN-13: 9781554891054

Published in Canada and printed simultaneously in the United States of America and the United Kingdom.

CIP data available from the publisher.

Table of Contents

Acknowledgements

I would like to extend my sincerest thanks to the following people who helped me turn this book into a reality. First of all, I want to mention that I took my preparation class for the Project Management Professional (PMP) certification examination from the RMC (Rita Mulcahy) organization. I distinctly remember my instructor, Jeff, warning me not to argue with the materials that were presented if I wanted to pass the certification examination. This was a real challenge for me since, unlike so many others in my class, I had been practicing Project Management for many years prior to this time and had formed certain impressions which differed with some of the material that was being shared with us. I especially took exception to certain human resource management comments made in the *PMBoK Guide*® (PMI's *Guide to the Project Management Body of Knowledge*®) that seemed entirely wrong to me, based upon my many years of project experiences. Jeff said two very important things to me about this situation. First, my experience could likely be incorrect since I had not previously been professionally trained as a project manager but instead grew into the position

5

as so many others do over a period of time. Further, he strongly recommended that I accept and study the material at hand as it was presented, pass the PMP examination and gain my certification, and then go on to challenge the material that I was questioning. This book is meant to further this very good recommendation from more than four years ago.

Second, I need to thank the various professional consulting firms, for whom I have been employed, for allowing me the great opportunity to participate in and lead their clients' ERP system projects. In addition, my membership and participation in my local Project Management Institute (PMI) chapter allowed me the opportunity to test run a short slide show presentation of this material at one of their monthly luncheon meetings. An overwhelming number of the attendees at this event from several years ago fully agreed with my premise, findings, and recommended solutions.

Finally, and most importantly, without the support of my wife and children, this good work simply would not have been at all possible to develop. Given the extent of my research and writing efforts, a good deal of time was devoted to this endeavor. My wife, in particular, allowed me to take all of the necessary time to create this book and for that I am and will forever be most grateful. My children greatly encouraged me to persist with this effort and to see it through to a successful resolution. I am most grateful that they all urged me to keep at it. My real hope is that my profession embraces this information to help make the practice of project management even more successful going forward.

I plan to continue to participate in this effort as I develop my competence as a project manager over many upcoming projects.

Richard C. Bernheim, PMP, MBA
February 2011

CHAPTER ONE

Introduction

This book came about as a direct result of two key events in my professional life. The first of these was the completion of nearly twenty years as an Enterprise Resource Planning (ERP) software system implementation consultant, team leader, and project manager. The second was more unexpected as a result of the personally-initiated research which I conducted subsequent to becoming a certified Project Management Professional (PMP) from the Project Management Institute (PMI). You see many of the experiences I had during those nearly twenty years provided me with a front row seat into the very challenging world of information technology Project Management. As an active participant playing in these various roles, I witnessed the majority of the twenty three ERP projects in which I was involved as not being very successful contrary to both my own efforts and that of a good many other colleagues, not meeting our clients' expectations. My subsequent research on the subject, much to my surprise, unfortunately only validated the fact that my information technology project experiences were not an anomaly, but were actually rather typical. I learned the following two key statistics from a large variety of researchers who studied the rather high rate of information technology project failures over the last several decades:

- There are three categories of information technology project issues: people, Process, and Technological. According to the researchers, on average, 80% of all project issues and challenges are directly attributable to people.

- Project managers spend, on average, 70% of their very precious time on non-value added project activities.

Given these rather startling, significant, compelling, and most disturbing facts, both the idea and need for writing this book came to me. As a project manager working in the field of information technology project management for so many years, and having had so many varied project experiences in which I was directly involved and which turned out so poorly, I became determined to share some of my experiences to bring some reality to these rather troubling research results. By sharing three distinct stories in Part One of this book, which are presented in the form of plays with scenes, I first seek to illustrate these study outcomes. My point in presenting this first section of the book in this particular format is to emphasize the significance and reality of these most disturbing findings. It is not my intent to share all of the specific details of these cases, such as which ERP software system was involved, which companies and consulting firms were involved, where the events took place, when the events occurred, and what transpired after sharing these events; rather, by ignoring these unimportant and tangential details, I seek only to reveal and concentrate on how these project situations unfolded and what, if anything, was actually achieved of any real and lasting value in the final analysis. By telling these stories as plays, while totally de-emphasizing such specific details, I believe I can share what is most important for members of the profession to glean from this book, while hopefully having a rather enjoyable experience in doing so.

Projects require time tested methods and tools for accurate, complete, and consistent planning and execution by all involved. Technology is often the driver for many information technology projects as both computer hardware and software continually get better and more advanced but methods, tools, and technology alone will not make projects get done well – you need people for that to happen. Without people projects simply could never get accomplished. Yet it is people who cause the overwhelming number of issues and challenges and who suck up a very high proportion of a project manager's time and energy over matters having nothing at all to do with methods, tools, and technology. People far wiser than I have said that project plans don't sink projects. Technology, as challenging as it is, doesn't cause projects to fail either but, instead, people do. Despite all of the many positive advances in the last several decades in methods, tools, and technology, the success rate of information technology projects remains well below a 50% breakeven point. The main culprit is very clear – it is people issues that cause projects to fail or, even more frequently, to achieve far less than was sought or desired by management. The challenges people generate come from an almost endless supply of reasons:

- Poor interpersonal skills and relationships – mainly having to do with a lack of communications (in listening, speaking, along with writing)

- Bad attitudes

- Fear of change

- Loss of power

- Lack of understanding

- The self-fulfilling prophecy

- Low confidence

- Cultural differences

- Generational conflicts

- Men versus women – their different viewpoints and approaches to project activities and decisions

- Racial bias

- Lack of commitment and interest

- Poor or inadequate leadership

- Lack of clarity and direction

- Not adhering to pre-determined processes

- Inconsistent communications – both those sent and received

- Lack of teamwork

- Etc.

People lead, perform, and support projects. They must put aside their individual likes and dislikes and work together closely as opposed to separately or individually. When this takes place, things tend to go very well with the final project outcome and everyone involved comes away from the experience feeling good. Unfortunately, such occasions are too few in number relative to the vast and growing number of information technology projects that organizations engage in globally every year. The three plays presented in Part One share both the good and bad aspects of people issues and challenges faced in real life project situations.

Following a short summary where I analyze the lessons learned from these three project plays, I offer the reader a practical set of solutions in Part Two of this book dealing with the reality of large numbers of information technology project failures. It simply would not be fair to raise such a concern without offering up a workable set of solutions. The antidote that I reveal in Part Two is clearly all about project eadership. Achieving this in the face of the many challenges stated above is by no means a given. But there are ways of applying the concepts I will share with you, which are learned, understood, and applied properly and then followed consistently will most definitely increase the chances for information technology project success instead of failure or mediocrity. None of this comes easily nor quickly, and this is certainly not foolproof by any means, but it is well worth a project manager's serious consideration effort given the substantial amount of time and resources put into information technology projects by so many people and organizations worldwide.

Part One:

Three Project Plays

CHAPTER TWO

Case One:
Project No-Go

In the Beginning

Once upon a time, there was a multi-national
company who had an operating division that was
struggling to survive. This division's customers
were simply livid about the fact that this division's
customer service representatives (CSRs) were
regularly unable to inform them of the status of their
placed orders. Customers would call in and discover
that these CSRs had no readily available or reliable
source upon which they could provide the current
status of a placed order. These CSRs would take
down the information pertaining to a customer's
request and then shortly thereafter return each call
assuming they were able to determine an accurate
current order status to begin with. Far too often,
when they returned a customer's call, all they could
offer up in the way of any useful information was
that the order had indeed been placed and that
it was now definitely in process. Sometimes, the
CSRs didn't even bother to return a customer's call
because they were simply unable to provide any

useful information of any real value to the customer. So, as a result, over time customers began to take their business elsewhere. Given this situation, this division's revenues and profits diminished with the passing of time. The division president, Jeffrey, was under significant pressure from corporate management to change the direction of the level of customer dissatisfaction, along with the corresponding negative impact this caused to both sales and profitability. The pressure that Jeffrey faced was both personally directed at the position he held in the company and to the entire division itself. Corporate management was now giving very serious consideration to either selling or discontinuing division operations unless conditions rather quickly changed for the better.

Separately and unrelated to this situation, this company was seriously considering the acquisition of a new Enterprise Resource Planning (ERP) software system. Jeffrey was among one of many executives who was part of a committee charged with performing all necessary due diligence regarding this rather important new company initiative. One of the many facts he gleaned from his participation in the numerous ERP software system meetings and demonstrations from the vendor was that, by employing this new ERP software system, customer service levels could dramatically improve because all customer orders would be continually tracked once they were placed into the system. This new ERP software system operated in real time, enabling all customer orders to be easily identified with a precise current status at any given point in time. So, after some careful consideration, Jeffrey decided to put forth an argument for his division to acquire this new ERP software system. He developed a detailed business case and then he presented it to his fellow executives. He asked them to consider allowing his division to serve as a pilot site for the company's implementation of the new ERP system. In so doing, the company could better test out the features and functions of this new ERP system at a minimal initial cost while at the very same time enabling his division to benefit from improved customer service levels. So, shortly thereafter, corporate management agreed to allow Jeffrey to proceed with this action plan and a temporary agreement was struck with the vendor of this new ERP software system for a limited number of user licenses over a specific period of time. Thus, the die was cast and the project was inaugurated.

With renewed life and a sense of purpose, Jeffrey immediately launched into his new mission with a vigor not seen in him for a very long time. His first order of business was to select a project manager for this new and exciting endeavor. After a little thought, he decided upon not one, but two people: Robert, a senior business manager; and Thomas, a senior information technology manager. Jeffrey strongly felt this combination would best satisfy this rather demanding and critically-important situation. The two men selected had a combined fifty-three years of experience with the company. Robert had actually started at the inception of this division. Thomas joined just a couple of years thereafter. They had worked together on various projects at this division several times before. By combining their knowledge and experience through co-managing this endeavor, Jeffrey felt that he had the ideal leadership in place to insure a successful and timely project outcome. Once they were named, both Robert and Thomas were relieved of many of their existing job responsibilities so that they could focus upon their new mission just given to them by Jeffrey.

Robert and Thomas now began to select project team members representing both the business and information technology aspects of this new project. They also acquired some initial training on the new ERP software system's features and functions that were relevant to the specific scope of this effort. The key modules of the ERP software system determined to be in the scope of this project were financial and cost accounting, materials management including inventory control, and sales and distribution. As the project team was forming, so too was the project plan. Further, given the fact that this company had no prior knowledge and experience with ERP systems nor the new client-server hardware technology associated with it, a search for a reliable consulting firm was commenced. In due course, a "Big 6" firm was selected to work with the project team on this important effort.

Now that a complete division wide project team was selected, including all necessary business and information technology subject matter experts, a detailed project scope and plan were developed, and the consulting partner's staff were added to the project team. It was finally time to get things underway. In short order, a project kick-off event was conducted where a good many details about this endeavor were shared with everyone involved. In addition to Robert and Thomas leading this three hour presentation, Jeffrey was asked to add his own encouraging comments about the importance of this new and rather exciting mission for both his division and the company overall. He made it quite clear that the primary scope of this project was about dramatically improving customer service by being able to continually track all customer orders from the time of their placement through to their shipment. Jeffrey emphasized that the new ERP software system offered his division and the company much more than just that, and probably, in time, such additional features and functions would likely be implemented. However, in the meantime, the scope of this project was both clearly set and firmly fixed.

The project was properly and officially launched. Robert and Thomas got the project team together to go over the project plan in detail, particularly regarding the near term activities. The first phase of the project, called Project Preparation, required the development of a project charter amongst several other important deliverables. Everyone got involved in contributing their good ideas to this effort. Led by Robert and Thomas, plus the project manager from the "Big 6" consulting firm, matters such as how to handle scope changes, issues and decisions, risks, project communications and status reporting, and project deliverable quality standards were dispensed with in fairly short order and with minimal dissention amongst all of the project team members directly involved. Things at this early point in the project lifecycle seemed to bode rather well for this important endeavor. Jeffrey was pleased and only hoped it would continue as such in the weeks ahead.

Making Progress

The second phase of the project, called Design, began with the current state of all relevant in scope business processes being explained in detail by the subject matter experts on the project team to the outside consultants in order to establish a baseline of everyone's understanding. Various "as-is" sessions were scheduled and then held to go over just how these specific business processes were currently being addressed by divisional end users. Once again, everyone involved went about the business of getting this activity done on schedule, and to the required level of detail and quality. All was going well and everyone was beginning to feel quite comfortable with the mission and working along with each other on it.

Once the current state was completely documented and verified, the project team turned its attention to the more demanding task of addressing the future state of business processes, taking into account the usage of the new ERP software system's features and functions. This required conducting ""To-Be"" workshops for each specific business process within scope and having a reasonably good understanding of the new ERP software system. This effort further required the subject matter experts on the project team to first consider new ways of performing these business processes. Then they needed to analyze and evaluate the various options offered by the new ERP software system, which represented "best practices", prior to recommending the one they all thought would be best for their organization going forward. All involved project team members took this important effort very seriously as they deliberated over the details of each business process while devoting adequate time to this most critical project activity. These workshops were called brainstorming sessions and they were conducted in a rather structured and consistent manner, guided along by the external consultants who had done this many times before in many other client project situations. By way of an example, one of these many "To-Be" workshops, and one of the first ones held, concentrated upon the sales order entry business process. Mary, a subject matter expert with a very substantial amount of knowledge and experience in all sales and distribution business processes at this division helped to drive the conversation and direction of this particular workshop. She contributed numerous good ideas for her fellow workshop attendees to ponder over. In addition, along the way, she communicated her personal life goal of emulating Martha Stewart whom she held in very high regard. This group regularly heard from Mary about just how much she admired Martha Stewart and the tremendous professional and personal success she had achieved. What Mary really wanted to do more than anything else was to be Martha Stewart and get as far away from her current job as possible, as soon as possible. Despite the annoyance of others on her sub-team, Mary performed her given tasks quite well. Her assigned consulting peer, Robin, helped document the many thoughtful and carefully considered ideas from all of the workshop participants, including Mary, for making this business process much better by taking into account their understanding of the features and functions in the new ERP software system that were specific to the sales order entry business process.

This entire group also took the time, with Robin's guidance, to research best practices pertaining to the sales order entry business process. They incorporated many of these ideas into their specific recommendations. Mary and her entire workshop group were very proud of their results. Once it was all fully documented in the appropriate business process template provided by the respective "Big 6" consulting firm, the document was routed to both Robert and Thomas for their review, and to seek their approval as was required.

After a short while, Robert and Thomas sought out Mary and her workshop group to give them all feedback on this particular business process document. Mary was feeling rather confident that they would hear positive comments from these two co-project managers. As the meeting was called to order, Robert, representing the business perspective, spoke first. He went on with a very harsh critique of this group's first significant work product. Robert attacked each and every recommendation that was made by declaring why it would not work at this division. When he was finished, Thomas proceeded to do the very same thing, just from an information technology perspective. When done sharing their rather biting remarks, the two co-project managers got up and abruptly left the room. Mary and all of the others were totally stunned by what had just transpired in less than ten minutes. None of them were able to comprehend what had just occurred, yet alone why it did. As they sat in disbelief trying to absorb what had just happened, Mary decided to speak up. She said that she was most surprised, but she would seek out both Robert and Thomas on her own to once again try and convince them of the merits of their many good recommendations. Everyone in the room agreed that this was best since no one else had any better suggestion at the time.

In due course, Mary got her audience with both Robert and Thomas in their project management office to restate the case for the ideas expressed in this future state sales order entry business process document. Yet again, both Robert and Thomas berated her and her group for putting forth such poor recommendations that would never work at this division and that Mary, in particular, should know much better than the others that this was the case. They were both most annoyed that Mary had once again taken up their valuable time in revisiting this same document that they had deemed entirely inadequate and inappropriate to the situation at hand, in their collective judgment. Mary was told yet again to return to her workshop group and to rework this business process by relying more upon current practices as their guide. She was also reminded that time was now of the essence in terms of this specific task. So Mary left their office feeling rather rejected and most disappointed that she was unable to convince these two co-project managers of the value of the entire group's many good and worthy recommendations, many of which were deemed to be best practices.

Mary gathered her workshop group together for another try, knowing full well that none of them was really interested in applying current practices to this business process, which they all felt was rather poor to begin with. They also acknowledged that time was now in rather short supply as they had to move ahead with the next main activity of this project. They met and discussed again the best ways of improving the sales order entry business process. In their earlier attempt, they had thoroughly dissected all of the inefficiencies and non-value added work steps in the current state process. They had taken into consideration not only their own thoughts and opinions on this subject matter but also real customer experiences and all of the negative feedback received about the current sales order entry business process at this division. As they reconsidered yet again all of this information and reread their future state business process document, they became even more determined that the recommendations they had previously expressed were both quite valid and certainly most appropriate for this division. Other than making some rather minor wording adjustments to their first effort, they resubmitted the document back to both Robert and Thomas. Within a couple of days, Mary was summoned by Robert. He berated her for putting forth the same ideas and approaches as before, despite what she had been explicitly told twice before. Robert further indicated that Thomas was equally disturbed by this situation, but he was too busy right then to participate in this meeting with Mary. Robert then proceeded to remind Mary that this important deliverable was now overdue according to the project plan. Given this fact, not to mention her and her team's insubordination, he told Mary to take the business process document describing the current sales order entry process and to mark it up to reflect the "To-Be" process by modifying the steps impacted by using the new ERP software system's specific features. Mary argued with Robert to no avail and she soon left his office

feeling very angry. Upon calling her workshop
group back together to share the latest news, she
started her remarks by talking about leaving the
company to follow in the footsteps of her beloved
Martha Stewart. Once she related her meeting with
Robert, and after the group had some time to take
in the full measure of what Mary shared with them,
they turned to young Robin, their group's assigned
outside consultant, for both some good advice and
support. Robin also indicated her disappointment in
the group's inability to convince Robert and Thomas
of the many merits of their recommendations but
she reaffirmed Robert's understanding of the project
timeline for this important task and then she told
the group to give Robert exactly what he had just
asked Mary to do.

Stormy Seas

Mary's group in its various good faith efforts
in addressing the sales order entry business
process was not alone in obtaining a rejection
of its recommendations from both Robert and
Thomas. Almost every other workshop group that
focused upon a specific in-scope business process
encountered the very same reaction to their own
proposals from these two co-project managers. As
word of this spread amongst the various subject
matter experts, and then across the entire project
team, open hostility towards Robert's and Thomas'
authority began to break out. Each of the external
consultants assigned to a specific workshop group
and business process was pressed into taking sides
by either supporting the subject matter experts or
the two despised co-project managers. This situation

was not at all what the consultants had ever experienced before and they certainly did not enjoy having to pick a side in this brewing battle. So, just like the subject matter experts, these consultants got together amongst themselves and with their own leadership to share what was happening, as well as what was now being asked of them to do. They all felt very strongly that the subject matter experts, along with their workshop group members, had put forth substantial time and great effort into this critical activity. Further, they expressed the fact that the many recommendations expressed in the future state business process documents were appropriate for this division, company, and industry. Finally, they all said that they simply did not understand at all why both Robert and Thomas rejected practically all of these recommendations.

Mary was so upset with the situation that she decided to take it upon herself to do some investigation into the backgrounds of both Robert and Thomas. She had never worked directly with either of them before but she had heard rumors over the years about them from other employees. After talking to several fellow co-workers in the company who had prior and direct contact with these two managers, she returned to her sales and distribution business process sub-team to share with them all of what she had discovered in this self-initiated effort of hers. What she revealed to her group was that Robert and Thomas had a long history within this company of going out together for alcohol-laden lunches, and for chasing after various young and attractive female employees who they could then manipulate for their own personal sexual advantage. Mary's sub-team, including their outside consultant

Robin, all became enraged by this news. Once their ad hoc meeting broke up, it wasn't very long at all before word of Mary's discovery spread throughout the entire project team. Some of the other sub-teams were still engaged in their specific future state business process workshops.

Soon thereafter, the entire project team, minus the two co-project managers, decided to meet to discuss how to handle this situation. They were all very frustrated with the reactions they each received to-date from Robert and Thomas to their many good recommendations and now this news only further upset them. It was now quite apparent to everyone that both Robert and Thomas were not at all interested in changing, yet alone improving, the current business environment. It was then decided that a couple of the more senior subject matter experts on the project team, along with the consulting project manager, would seek a meeting with Jeffrey, the division president, just as soon as possible to explain what had been happening and why the project had now fallen somewhat behind schedule at this early stage in the project. They would further request that Jeffrey review a sample of the future state business process documents already rejected by Robert and Thomas to determine his reaction to them. They soon discovered that Jeffrey was out of town fighting one of the many fires existing with his division's customers, but that he would return in a few days to address this urgent request.

Upon his return, Jeffrey hastily called the meeting requested by the selected subject matter experts along with the consulting project manager.

He immediately told the group that he had only about half the time that they required of him that day. The respective subject matter experts then decided to have Mary share her story by first explaining to Jeffrey her workshop group's specific future state sales order entry business process recommendations, and then how these were received by both Robert and Thomas. After hearing from Mary, Jeffrey requested from the other subject matter experts their respective future state business process documents so he could read them on the airplane ride he was taking later on that day to visit yet another unhappy major customer. They agreed to get these to Jeffrey prior to his departure. Jeffrey finally added that he would not return to the office for a few days, and that the group needed to be patient until then. He promised to thoroughly review all of these detailed documents and to reconvene this group just as soon as he could upon his return to the office.

In the interim, Robert and Thomas caught wind of the group's recent meeting with Jeffrey. They too became rather upset with the fact that their specific orders to adjust the "As-Is" environment by taking into account specific features and functions of the new ERP software system were still being totally disregarded by the entire project team. They decided to take out their wrath upon the consultants who took the side of these rebellious subject matter experts. As a result, everyone involved with this project was both angry and upset with both Robert and Thomas who seemed not to care one bit about this in the least. However, some project team members did notice that their daily out of the office lunch periods became much longer, and that when they returned to work they both acted rather differently in the afternoons than they did in the mornings.

Upon Jeffrey's return, he quickly summoned the subject matter experts and the consulting project manager he had met with earlier along with both Robert and Thomas. He told them all that, for the most part, these documented future state business process recommendations were quite good and they met with his approval. Robert and Thomas acted very subdued throughout this encounter with Jeffrey and the rest of the project group. After the discussion broke up, they both requested additional time with Jeffrey. He granted them just a very few minutes. Robert and Thomas proceeded to inform Jeffrey that the subject matter experts and consultants alike were uncooperative and antagonistic to them and to their authority over the project. Jeffrey responded by telling Robert and Thomas that they would both need to work much harder at building a better relationship with these subject matter experts and external consultants. Then Jeffrey raced off to yet another important meeting he had elsewhere in the building.

The subject matter experts led by Mary felt exonerated as they proceeded to convey to the rest of the project team and outside consultants what had just transpired with Jeffrey. The consultants were convinced that the project would now get back on track and be able to make further progress. Unfortunately, this would prove to be wishful thinking; Robert and Thomas were not about to accept defeat and give in to this rebellious group of people. Instead, they discussed how to win their upcoming battles and reassert their authority over the entire project team, including the external consultants. Following their own conversation, they proceeded to go out for a rather long liquid lunch.

The future state business process procedures were now finalized and resent to Robert and Thomas who reluctantly approved them. The "To-Be" process documents were approved by Jeffrey and were all subsequently returned back to the subject matter experts to use going forward. The next phase of the project involved configuring the new ERP software system to achieve the many detailed goals defined in the now approved "To-Be" business processes. During this part of the project, the consultants sit side by side with their fellow project team members in order to configure the software and prepare the hardware according to the details spelled out in the future state business process design documents. In so doing this, knowledge would begin to transfer from the consultant to the designated company person. Given all of the features and functions of this particular ERP software system, it was common for the consultants to explain the various alternatives offered by this system to accomplish the desired business process result. As a result, project team members often needed to check in with their respective subject matter experts for specific guidance and direction. Sometimes such conversations were both quick and painless, but too often this was not the case. Subject matter experts had no choice but to make a case for one option over other possible alternatives, and then present this information to both Robert and Thomas in order to obtain a final decision so the project work could proceed. Given their recent experiences with

2 - Case One: Project No-Go

these two co-project managers, the subject matter experts dreaded having to take such matters to Robert and Thomas. On the other hand, Robert and Thomas viewed such encounters as opportunities for payback. To them, this process was just a game to be won over and over again each and every day. Furthermore, they knew that such decisions would not require Jeffrey's involvement. As you would expect, Mary was the first person to bring them such a matter. Mary had her situation well described in writing, laying out the three options offered by the new ERP software system. For each alternative, she stated all of the pros and cons. Finally, she took a position by stating a recommendation along with the various reasons as to why this was best. This particular matter was rather complex since it had both functional and technical aspects to consider. Mary turned over her document to Robert and Thomas asking them when she would have their answer. They both told Mary it could take awhile because they had more pressing matters to attend to, including taking additional classes on the new ERP software system at a location that was out of town. Mary was disappointed but had no choice other than to accept their response and to just wait it out.

The resolution to this situation impacted several other business process configurations. Time passed, and Mary tried to be patient, but her consultant Robin was pushing her for a result as this matter was now beginning to impact overall project progress and team momentum, not to mention that it would soon begin to reflect badly upon the both of them. Finally, it was Thomas who got back to her. He called Mary to the project management office and then proceeded to explain to her why another option,

and not the one she had recommended, was superior.
Thomas used his deep technical background to
explain his rationale, which completely overwhelmed
Mary. Thomas assumed Mary would have no choice
other than to accept his argument since she was
a business person, not an information technology
person. He underestimated Mary yet again. She
listened intently to Thomas's explanation and took
very detailed notes since she too realized that she
was not in a position to challenge Thomas. Mary
then visited Robin to convey this decision and the
rationale expressed by Thomas in support of it.
Robin, while younger than Thomas, had a solid
information technology background along with
some reasonably good business experience. Robin
declared Thomas to be completely wrong on this
important matter. She insisted that they needed
to revisit the matter with Thomas as soon as
possible. At the appointed time, they both went to
see Thomas in the project management office armed
with Robin's written counterpoints. Robin tried to
explain her various points. Before she was able to
utter no more than three sentences, Thomas reacted
harshly. Robin, being much younger than both
Thomas and Mary, and definitely more respectful
than Thomas, got very quiet allowing Thomas the
added opportunity to further berate her by stating
that their current technical environment would not
be able to support any other option than the one he
previously communicated to Mary. Having been
on the project for several months, Robin should
definitely have been aware of this fact, implored
Thomas. Further, he was now very angry that more
precious project time was being wasted to bring this
closed matter up once more. Thomas then proceeded

to inform Mary that he would now inform Jeffrey that Mary and her sub-team were once again falling behind schedule and continuing to waste company funds. Then, he rather abruptly got up and walked out of the room, leaving both Mary and Robin in a state of shock.

This very same routine began to play out with several other subject matter experts. As word spread throughout the project team about this new situation with the two co-project managers, another full-scale rebellion began to develop. Both company staff and external consultants began to discuss what to do about this situation, which meant that little real project work, yet alone any progress, was now taking place. Soon enough, this became evident to both Robert and Thomas who proceeded to use this new issue to inform Jeffrey yet again of the entire project team's lack of effort and discipline. After a brief while, they managed to convince Jeffrey that, despite their very best efforts, the project team was refusing to cooperate with them on this critical project activity. As always, Jeffrey was very busy with more pressing matters having to do with either holding onto existing customers or fending off other corporate executives bent on closing or unloading this poorly-performing division. With customers, Jeffrey would tell them to hang in and await the implementation of the new ERP software system which he said would dramatically improve customer service levels. With his fellow executives, he would use the very same argument and remind each of them that his division was the pilot site for their own future use of this new system. So, given Jeffrey's daily struggles to survive, the very last thing he needed to learn from Robert and Thomas was how badly things were going with this critical project. He became extremely annoyed with the whole situation and began to wonder why he ever bothered with this entire endeavor to begin with.

A Category Five Hurricane

Jeffrey decided that he needed to pay an unannounced visit to the project "war room" to reiterate both the urgency and necessity of the design activities to the entire project team. As always, his time was very limited so he asked that everyone stand in a circle around him so he could deliver his brief remarks and possibly take a question or two before he had to run off. Twenty-three people gathered into a circle around Jeffrey in the middle of the war room to hear what he had to say. He reminded them of the criticality of this important project to the survival of this division. If the new ERP software system did not "go-live" in just four months time, there was simply no hope whatsoever for this division to survive intact. Jeffrey said that he was telling all existing customers and the other company executives to hold on until this new ERP software system became fully operational. Unless the project team delivered according to the approved project plan, there was no hope short of a miracle that the division's staff would keep their jobs with this company. Then, he turned his attention to the outside consultants and said that a good client reference from him and this company would not be forthcoming unless this endeavor was successful. This "Big 6" consulting firm badly wanted to expand its presence in the industry in which this company operated so a good reference resulting from the project was important to them. As well, any opportunity to implement this ERP software system throughout the rest of this large global organization would be lost. Then, Jeffrey checked his watch and noticed that this effort took a bit longer than he had planned so once again he raced from the war room to another appointment, leaving the audience speechless.

Robert and Thomas decided to use this opportunity to add their own words to what Jeffrey had just shared with them all. Robert went first and proceeded to say that this entire group needed to start acting once again like a team. Thomas interjected his thoughts on the matter, which basically mirrored what Robert had just said. Then, these two co-project managers left the room to head back to their office before going for lunch. Mary spoke to those still standing in the war room. She reminded them that it was not them but rather Robert and Thomas who had brought this entire situation upon themselves and everyone else. Mary said she was very tired of the constant stress that she and her sub-team continued to experience from Robert and Thomas, and that she was again considering taking early retirement to pursue her lifelong dream of becoming the next Martha Stewart. Everyone else wanted to complete this project, and the company staff definitely wanted to continue to learn much more about this new ERP system that was taking the marketplace by storm. Almost all of the company project team members had discovered just how valuable such knowledge and experience would be. Some of them personally knew people who had been exposed to this new software and then left their employers to take a position with another employer for significantly higher compensation. Opportunities for these team members to do the same were both plentiful and certainly not any kind of secret. This also included the opportunity to make even more money by joining the growing ranks of the "Big 6" consulting firms actively looking to hire such experienced professionals of this hot new ERP system. The stand-up meeting finally ended with everyone going back to their specific project activities feeling very down about this entire endeavor.

It was a new day. The night before, Mary had discussed the details of the situation at work with her husband, Jim. Now she returned to work determined to give it alone last try. She told her sales and distribution sub-team that she would request another private meeting with Jeffrey with whom she had worked with for over twelve years. Mary wanted him to know that Robert and Thomas were the source of all the friction for the entire project team since the inception of this effort. She wanted Jeffrey to know that all of the sub-teams and the external consultants were fully dedicated to the successful completion of this vital project. First, she arranged a meeting. Then, she headed off to this office to try a personal approach to what had now become a full-blown crisis for everyone involved. Jeffrey seemed to be in better spirits than usual when she entered his office. He told her that his fellow company executives had just agreed not to take any action until after the new ERP software system was productive for at least two months. This small victory sounded nice to Mary as well, but she had more pressing matters of her mind to share with him. First, she proceeded to remind Jeffrey of their past work together at this division and how she had helped him several times before with various challenging business and employee situations. He indicated that he also recalled these past work experiences and he agreed that Mary's help had proved to be invaluable. She then explained to Jeffrey what had transpired to-date with this project, noting that it had been Robert and Thomas against the rest of the project team, including the outside consultants. Mary went on to tell Jeffrey that it was also no secret that both Robert and Thomas regularly took long lunches at the local bar. She also mentioned the well-known fact that Robert's administrative assistant, Sharon, was having sex with him on the company premises during working hours in order to better secure her job in the company, and that she had recently obtained a hefty salary increase thanks primarily to Robert's review of her job performance. Mary reminded Jeffrey that Thomas had a very similar situation with his former administrative assistant, Jill, who had left the company just prior to the start of this project. After a long pause, Jeffrey then admitted to Mary that he was well aware of Robert's and Thomas' personal failings. He then proceeded to tell her that just before he announced their role to co-manage this new ERP software system project, he had told each of them that unless they shaped up, this opportunity would be their very last chance to stay with the company. At the time, they both swore to Jeffrey that they would change their ways and deliver for him. Jeffrey now realized that their bad behavior had gotten far worse.

At this stage in the project lifecycle, Jeffrey felt he had no choice other than to endure this situation until the project was delivered. Mary understood his rationale, but she certainly did not agree with the outcome that he just proposed to her. Down deep, she knew that both Robert and Thomas would continue to make life impossible for her and her fellow project team members for the remainder of the project. Mary thanked Jeffrey for his frankness, and then left his office feeling that all was lost. She returned to her sub-team and reported all that Jeffrey had shared with her. As usual, within a couple of hours the news spread to the rest of the team members. Everyone now shared Mary's view that the project was doomed.

This self-fulfilling prophecy would soon prove true. Joe, the financial and cost accounting expert on the team announced his departure from the company to take a position with another company that was just beginning to implement the same ERP system. Within three weeks, two more project team members announced their voluntary departures to take similar work. After a little more time, three others would follow along the very same path. It was clear to everyone that the project could no longer effectively function so the rest of the team disbanded and the outside consultants were dismissed. The new ERP software system would never "go-live" at this division of this company. No one involved would ever really know what eventually happened to Jeffrey and his division.

CHAPTER THREE

Case Two: Project So-So

The Thrill is Back

A large, long standing, and successful company was undergoing a very substantial shift that was being forced upon them by purely external commercial factors. Big governments were no longer spending what they had previously on certain types of equipment and their respective replacement parts. As a result, this company realized it needed to chase after more commercial business, which meant modifying many of their existing product lines to fit into this new and growing base of customers, while curtailing many of their other operations. By so doing, a good many internal business processes would need to undergo definite and significant re-engineering. Information systems were also rather dated, not to mention that Y2K was fast approaching and the company was totally unprepared for it. The senior management team decided to select and then implement a new Enterprise Resource Planning (ERP) system with the assistance of a large consulting firm experienced in doing this within

41

the company's industry. A project was launched to accomplish the twin goals of substantial business process re-engineering to fit the new customer business environment, and implementing the new ERP system to act as the primary engine in support of the new businesses processes going forward. It had been a long time since this company had engaged in such a large scale information technology project. Given this fact, many managers requested to be part of this exciting new endeavor, making it take longer than usual to sort out all of the key players for this effort after the project manager was selected, announced, and then put into position.

The rest of the company's project team members were selected and they totaled eighty five people. All of the subject matter experts representing specific modules of the new ERP software system were men who had each been with this company for more than twenty years. As such, their experience was when the focus was on the big government customers and not many smaller commercial ones. Each of these men had served in some kind of governmental capacity prior to joining the company. They all knew each other very well having worked together for a very long period of time. These men were all "can do" Type A personalities.

John, the project manager, would make sure that the positive attitude of the subject matter experts would be applied to every aspect of the new ERP system implementation project: developing the project plan, how they would handle issues and decisions, setting quality standards for the project team's deliverables, handling the project team members and the outside consultants, etc.

John was shorter than most of the other men and bore a distinct resemblance to Napoleon Bonaparte, whom, it turned out, was one of his heroes. John had been with this company for over twenty five years in a large variety of capacities and with his business and company knowledge was well prepared for all aspects of this important project. However, he had never managed an information technology project, nor a re-engineering project, ever before. This was true for all of the other four company subject matter experts as well. They felt that the consulting firm chosen to work on this endeavor would compensate for whatever the company project team managers were lacking. The team members were oriented to their new mission with a day-long project kick-off event led by John. Everyone was excited about this new initiative, which clearly represented the future of the company. While some of them had been involved in other company improvement efforts over the years, nothing of this magnitude had ever been tackled before. Process re-engineering and implementing a new ERP software system was daunting to them all. No one expected the challenge to be easy, but it was certainly deemed to be achievable with good effort.

Following the project kick-off, the real business of getting the project underway began. The first major activity to be addressed on the project plan was establishing detailed documentation regarding the current state business processes that were within the scope of the project. The company had never maintained business process procedure documentation previously, only preparing such documents when a government contract called for it. Even these limited number of business process

procedures were out of date. The consulting firm
had named two co-project managers to work with
this client. George was an expert in business process
re-engineering with knowledge and experience in
the industry within which this company operated.
Brian was an expert on the new ERP software to
be implemented. Since the first major activity had
to do with business processes, it fell to George to
establish exactly what was now required from the
project team. George developed the many specific
tasks in the project plan for this activity to go
forward. He had worked closely with John and
the company subject matter experts on developing
all of the details around conducting the sessions
to document the current state business processes
including the due dates, the review process to be
adhered to, and the necessary quality standards.

The "As-Is" process documentation sessions got
underway. There were four sub-teams working
on financial and cost accounting; materials
management, including inventory control; sales
and distribution; and production planning and
control. In addition, there was a technical sub-
team dedicated to handling all of the hardware and
software aspects of the project. All of these sub-
teams were now charged with gathering current
state information of the business processes and
technical environment. These sessions were to share
this detailed information and to begin to document
it into the standard templates provided by George.
Given that this was all about the current state, a
somewhat limited timeframe was placed around
this first major project activity. After all, it was
expected that things would dramatically change
in the future state, so why spend any more time

than was necessary to document the current state of affairs. George reminded everyone that this exercise was important for another less obvious reason: by detailing the current business processes and technical operating environment, a baseline of performance could be established for future comparisons after the project was completed. He told everyone not just to create flowcharts in detail with explanatory words, but to gather current volume statistics including the number of specific activities performed, the number of people involved, and the level of internal and external customer service achieved. The other consultants provided guidance to the company project team members to help them better understand these various detailed requirements. These tasks proceeded smoothly and on schedule under George's direction. Everyone felt the results achieved were excellent, which was cause for a small celebration after five weeks of their collaborative efforts.

Onward and Upward

The next significant project activity was to address the future state business processes along with the technical environment in which they would soon operate; doing so by taking into account the many features and functions of the new ERP software system. George gathered the entire project team together to explain the various important tasks involved with this critical activity upon which much of the ultimate success for the project would hinge. He told them all he had scheduled Brainstorming workshops in the project plan around each and every business process and technical aspect within

the scope of the endeavor. George went on to say
that unlike the effort already put into the "As-Is"
sessions, the "To-Be" workshops would be far more
mentally demanding upon all of them. Previously,
all they had to do was determine and then document
the current state, and then have their flowcharts
and verbiage verified and approved by the respective
business process owners. This time, they would need
to ponder every conceivable way to dramatically
improve these business processes and technical
aspects. The external consultants would interject
ideas into these brainstorming workshops based
upon their knowledge and experience, plus everyone
would need to explore and seriously consider "best"
practices. George told them all to focus intensely
upon this effort so that their final results would
reflect well upon themselves and the future of their
company. He said that this entire activity was
allotted ten weeks in the project plan and that about
sixty percent of this total time should be devoted to
generating ideas while the remaining forty percent
of this time would serve for documenting their
work efforts, and obtaining all necessary business
process owner and executive management approvals.
George then charged the project team with two
last thoughts before sending them on their way to
begin this vitally significant project activity. The
first was a set of ground rules to be followed for
each "To-Be" workshop as they went about the hard
work of generating better ways of performing the in
scope business processes. His list of ground rules
consisted of such matters as not putting other fellow
project team member's ideas down, not interrupting
people, not attacking people, and remembering to
turn off their cell phones and Blackberries. The

second and final piece of advice George shared with
them all was that they needed to be well prepared
to defend all of their recommendations when the
time came to do so. He suggested that one of the
best ways to do this was to quantify process work
steps and then contrast these expected results with
present circumstances whenever it was possible to
do so. For instance, if a business process currently
had eight distinct work steps from start to finish
while the new and improved business process
had just five work steps, then this would clearly
demonstrate to senior management that the process
going forward would likely require a much shorter
duration, probably resulting in lower costs with
far fewer opportunities for errors to be introduced
into the business process. Finally, they were asked
to cite other leaders in their industry employing
these new and improved business processes; more
importantly, to cite those achieving good solid
results from them.

The project team was dismissed to begin their
critically significant work. They went off to tackle
their new project mission with both enthusiasm
and great interest since none of them, except for the
outside consultants, had ever had such a chance to
do anything like this before. They were all excited
about what they would learn and accomplish from
participating in this rather challenging effort. Now
the "To-Be" workshops got underway in earnest.
One of George's ground rules was not to allow any
outside interruptions to interfere with their "To-Be"
workshop activities. This message was shared by
John, the company project manager, with all other
company managers who had staff involved in this
project, and especially those directly involved in

these "To-Be" workshops. Despite this, the company Controller, Ronald, interrupted a workshop to pull out Peter, who was a member of the financial and cost accounting sub-team, for two hours to assist him with year-end staff performance evaluations. Peter's absence came during a very lively and intense discussion his workshop group was having at that particular point in time. Further, it disrupted the group's momentum from which they did not recover on that particular day. Of course, the consultant involved in this workshop group, Susan, reported this event to George who, in turn, shared the news with John. John's reaction, after he spoke with Ronald was that there was nothing he could do about the matter but he informed Ronald to avoid doing this again or, at the very least, wait until these future state workshops were to be completed in six and a half weeks' time. Ronald was completely unfazed by this and a week later pulled Peter out of another "To-Be" workshop for a couple of hours to assist him with finalizing the company budget for the upcoming year. Peter was not happy about this situation but he really had no choice in the matter. Both of his responsibilities, to the project and to his daily financial duties, were very important as well. Word got back to George and, once again, he immediately went to complain to John about this matter. This time, John told George very firmly that there was absolutely nothing more that he could or would do about this situation, and that George and the others should expect it to likely occur again before the "To-Be" workshops were finished. Indeed it did, several more times, also involving other company managers and their respective staff who were directly involved in these workshops. Despite these interruptions, the effort moved forward as best it could under the circumstances and achieved reasonably good results.

Getting the workshops done and the detailed requirements fully documented on schedule was a significant achievement given the number of people, workshops, and complex business processes that were involved in this exercise. However, the hardest part of this important task was yet to come, which would be getting the approximately two hundred recommendations across all four of the business processes (and the technical project sub-teams) approved by John and the other subject matter experts prior to their formal presentation to the respective business process owners and executive management. The first point of review and approval was with the four subject matter experts who represented the four major business process areas within the scope of this project. One of these men, Andrew, the financial and cost accounting subject matter expert, had the closest and longest relationship with John. They both started working at this company around the same time and they had worked together over the many intervening years on other important company initiatives. Andrew's functional area produced some seventy recommendations that now needed to be evaluated and considered for his approval. The time to do so was short, relative to the time required to give this effort proper justice. So Andrew selected a sample of twelve recommendations covering the more fundamental financial and cost accounting business processes that would certainly get the attention of the Chief Financial Officer and Controller who were his key business stakeholders. Working overtime to read and mark up these recommendations with his comments and questions, Andrew was now ready to share his overall impressions with John.

He reported to John that the majority of these ideas and recommendations had been investigated in prior years by various financial and cost accounting managers involved with other much smaller business process improvement initiatives and, for the most part, were found to be either unworkable or of far less value than was first thought. In summary, Andrew told John that there was nothing really new in these recommendations that wasn't already known or that hadn't been considered in recent years. John was surprised at this rather dismaying report from Andrew but he knew to trust his judgment. John also knew just how tight Andrew was with both the Chief Financial Officer and Controller. Ronald, the Controller, had already announced his intention to retire in a year which just happened to coincide with the planned "go-live" date for the new ERP software system. If all went well with the project, Andrew was clearly in line to replace Ronald at that point in time. John now decided to check in with the other three subject matter experts to see how their sub-teams' ideas and recommendations were coming along. Before too long, it became quite evident to John that the impression he had gotten from Andrew's analysis also applied to the other sub-teams as well. This finding disturbed John very much so he proceeded to pay George a visit.

George was in a very good mood. In terms of the project, he was regularly receiving both directly and indirectly, word about just how well the "To-

Be" workshops were going in terms of people contributing great ideas and good efforts, adhering to the project's schedule, and the high standard achieved in documenting future state business processes. In addition, he just got word from his wife back home that she was pregnant with their second child. To make matters even better, she told George that they were going to have a son. Their first child was a three-year-old daughter. George couldn't wait to tell someone else the big news when John showed up in his office. When John asked George how he was, George jumped at the opportunity to share his great news. John congratulated Andrew and then proceeded to share what was on his mind which of course had to do with the observations of his four subject matter experts on the work products resulting from the "To-Be" workshops. George at first was only half listening to John when suddenly he heard something that grabbed his attention. John used the word "worthless" to describe the future state business process deliverables. So George asked John for examples in support of his claim. John primarily relied upon Andrew's financial and cost accounting business processes in conveying this reaction to George. What John didn't realize was that George had just finished reading a large portion of the financial and cost accounting sub-team's business process documents and his impression was quite the opposite. The only thing these two could agree on this day was how terrific Andrew's news was regarding his wife's pregnancy. In terms of the project, they now held totally different points of view.

Trouble in Paradise

The "To-Be" project activities continued to move forward to closure. The company subject matter experts at John's direction made the project sub-team members revise their recommendations to reflect business processes much more similar to the current state. By doing this, the respective business process owners and executive management would soon go on to approve these "To-Be" business process scenarios. Of course, the project team members, especially those from the company, were extremely unhappy with this state of affairs and in their project leadership. Despite voicing both their disappointment and disagreement with making such changes to their well thought out recommendations, they quickly realized and certainly understood that John and the four subject matter experts had total and absolute control over this project and all that was produced by it. Without much enthusiasm, they approached the upcoming milestone celebration where John would thank them for their hard efforts and then urge them to keep it up for the next series of work activities in the project plan. This milestone event would be an all day affair since there was so much for John to cover. In order to have all eighty-five company staff, plus the external consultants and project managers gather in the very same place at the very same time, another company facility was required a few miles away from the project "war room" location.

On the day of the milestone celebration, everyone shuffled over to the large auditorium where this daylong event was going to take place. John was the primary speaker along with Brian, who was George's co-project manager. Brian, up to this point in the project, had a role in support of George's re-engineering efforts but now it was time to focus more upon the implementation of the new ERP software system, which put the project spotlight on Brian from this point forward. Brian was certainly well aware of the challenges of dealing with this client, particularly John and the four company subject matter experts. He had supported everything that George and the other consultants had tried to do with the "To-Be" business processes. He also had his share of battles with most of them, including John, but he tried to put the best face he could on things for the sake of the project and his own employer, the "Big 6" consulting firm. Even what was to be said and how it was to be said at this milestone event had been the subject of many disagreements between John and his subject matter experts versus Brian and George who represented the outside consulting firm. As usual, John got most of everything he wanted. John and Brian spent the first part of the day summarizing the project activities and results achieved to date, especially pertaining to the recent completion of the future state business processes and the technical environment to support them all. The room was darkened in order for John's Power Point slides to stand out better on the big screen. John was up on the stage while making his remarks. Joseph, a more recent project team member from the "Big 6" consulting firm was also in attendance.

Joseph did not start this project at the very beginning like everyone else. It was not until near the end of the "As-Is" sessions that the need for his skill set was identified. Only then did the consulting team leader for the financial and cost accounting sub-team, with Andrew's agreement, launch into an

intensive search to identify a consulting resource
with the necessary knowledge and experience that
was now required going forward. Following this,
the identified consultant, Joseph, was interviewed
by telephone by both John and Andrew. He was
subsequently accepted and brought onto the project
team within a matter of days. Joseph got to spend
the better part of the "To-Be" effort working with
his sub-team on their many deliverables. Like
everyone else, he was well aware of the way John
and his fellow subject matter experts were running
the project. Anyway, during his phone interview
with John and Andrew, he told them both that he
was legally blind. What this actually meant in
terms of the project was that he could not operate
an automobile on his own accord but if they supplied
him with a seventeen inch computer monitor,
he could perform his project activities just fine.
This requirement was satisfied by John without
any further discussion and now Joseph had been
performing well in terms of the project up to this
point in time. Back at the milestone meeting, John
was delivering his presentation when he spotted
Joseph looking down at the floor from his seat near
the back of the darkened auditorium. Since Joseph
was unable to actually see the slides being projected
from so far away, he decided there was no point in
staring at something he could not see.

At the first morning break, John chased down
Brian and the financial and cost accounting
consulting team leader, Martin. John was quite
worked up as he told them both to dismiss Joseph
from the project immediately. A shocked Brian and
Martin asked John why. John quickly explained
to them both that Joseph had fallen asleep during

his remarks and such bad behavior simply would not be tolerated. Martin who sat diagonally behind Joseph during John's presentation, had noticed Joseph staring down at the floor and not up at the big screen but he also knew that Joseph had not fallen asleep but was rather intently listening to all of John's remarks. Joseph was summoned to an ad hoc meeting in a small conference room just off the main hallway outside this large auditorium. He confirmed exactly what Martin had just shared with both John and Brian. Then Andrew was asked to join this hastily assembled meeting. Upon joining in, he confirmed that what John said about Joseph was indeed the truth for he also witnessed the same behavior as John had, despite Martin's account of the situation. John then told the group his mind was made up and that Joseph had to leave the project right then and there. John then dashed out of the room to get in a quick bathroom break before the meeting reconvened for his continuing slide presentation. Andrew reiterated John's declaration. Then both Brian and Martin indicated to Andrew that by removing Joseph at this point in the project it would leave a hole in the project team that likely could not be filled for some period of time. Andrew said he didn't care and then he too dashed from the room to escape any further remarks from both Brian and Martin. Left behind in this small room were Joseph, Brian, and Martin, who were still reeling in shock from what had just transpired and were trying very hard to contain their growing anger. George then found them all in the room saying he just ran into Andrew and heard the awful news. He asked if there was anything he could do and Brian replied by saying "No, it's too late for that."

Joseph was put into a taxi cab to return to his hotel to pack his things and return to his home. That left George, Brian, and Martin to figure out how to handle the situation now that Joseph was gone. There was still nearly a full day of presentations ahead before a planned dinner celebration scheduled for that evening. Aside from a quick telephone call by George to the partner of this "Big 6" consulting firm to inform him of the events that had just taken place, they all proceeded to go through the balance of the day as best they possibly could. At the lunch break, several people began to ask about Joseph's whereabouts, especially those on his sub-team who worked very closely with him and knew about his disability, and who took extra care watching out for him because of it. Brian and Martin did not want to hide what had taken place earlier but they knew, on the other hand, that this was not the best time or place to get into the details of this matter. They said Joseph had to leave and they would explain the entire situation prior to the scheduled dinner. The afternoon went along as was planned as John and Brian explained the nature of the upcoming project activities.

Once the milestone meeting ended for the day, George and Brian gathered their fellow consultants into the very same meeting room where the terrible events of the morning had unfolded. They took turns informing them all of exactly what had taken place and the clear directive given to them by both John and Andrew regarding Joseph's forced departure from the project. Everyone got very quiet for several minutes taking in what they had just heard from George and Brian, before a brave person asked the first question. Susan asked Brian whether or not

John and Andrew understood that their sub-team's progress would definitely be impeded due to Joseph's sudden and potentially lengthy absence until a replacement was found. Brian said the matter was certainly shared with both John and Andrew and based upon their reactions it was obvious that they simply did not care. Then another consultant asked about how John and Andrew expected them, along with their fellow company project team members, to simply put this behind them and move forward with the project under these difficult and rather sudden disturbing circumstances. Brian said they simply would have to carry on but things would definitely not be the same as they were before this incident. It was clear to all that John would run the project as he liked and his four subject matter experts would back him no matter what really happened. The group became very melancholy and even more subdued. They really did not feel at all hungry for dinner, and certainly not up for celebrating with both John and Andrew. George reminded them that they should demonstrate unity with the rest of the team with whom they had already worked so hard and that they would need to continue to do so going forward. He said that he would understand if someone really did not want to join in for the upcoming dinner but that, as a whole, they really should do so. The group took a quick hand vote and decided to participate despite the fact that none of them really wanted to do so.

The evening came and went, with everyone having a very quiet dinner and doing their best to avoid both John and Andrew. The days following proved to be even more difficult for just about everyone except for John and Andrew, plus the other three company subject matter experts. The only thing John and Andrew now seemed to care about was determining exactly when Joseph's replacement would be ready to interview with the two of them. It took six weeks before a consulting resource became available, was interviewed, and finally was brought on to the project team. By that time, the pain of what had taken place was fading away, given all of the many current project activities that were well underway. The project team, especially the company members, went about their duties in a state of numbness, not expecting any support or appreciation from their project leaders. There was no point in suggesting any "best" practice since they would shoot them all down as unworkable or unnecessary. Also, there was no point in putting in extra effort since this too would not be recognized nor acknowledged. They just went about their project work activities without any feeling of real interest in the endeavor.

A Glass Half Empty

Now the project moved along into the software configuration and testing stages. Decisions were still required by John and his four subject matter experts but now they were just on a much smaller scale than before. Each and every time members of the project team attempted to put forth a recommendation, John and these subject matter experts found a reason to change it so that the outcome was more in line with current instead of any potentially improved conditions. The company project team members really wanted to demonstrate to their respective business process owners and executive managers that they still cared about what they were doing and that they wanted to better the company in the near future. When something came up that mattered a lot to them, a small group of them would pay a visit to the respective business process owner in order to obtain some positive feedback about their idea. Of course, once John and the respective subject matter expert found out about this, they made life even more difficult for the players who were directly involved in such a subversive action. There was growing tension between John and the four subject matter experts versus everyone else on the project team. The consultants, led by George and Brian were constantly fighting with the group led by John. Whether the issue was small or large, they always took the opposite viewpoint from what George and Brian attempted to put forth. Further, the hours and travel expenses incurred by all of the consultants got a great deal of attention by these company project leaders. John and the four subject matter experts began to challenge the number of hours they were being charged along with the travel expenses from all of the consultants. With each challenge came the need for a discussion involving George and Brian having to explain and defend the hours worked and the travel expenses incurred by each of their respective consultants, even including themselves. These meetings were quite hostile, often resulting in one party or another storming out of the meeting room. The consulting partner would visit on occasion, but the atmosphere was no better during his visits than it was otherwise. At one point, John specifically told Andrew to inform the Controller, Ronald, to put a hold on paying any of the outstanding invoices from this "Big 6" consulting firm until every last detail was fully resolved to his personal satisfaction. Of course, this made things all the more difficult for George and Brian.

In the meantime, within the financial and cost accounting sub-team, another critical project issue was now brewing. A disagreement was developing into something much worse. The matter had to do with a decision by Andrew regarding a software configuration matter. The company project team members on this sub-team along with their assigned consultant, Susan, had previously evaluated several alternatives allowed by the new ERP software system regarding a specific accounts receivable function having to do with customer account balance write-offs. This sub-team analyzed the pros and cons of each option and then recommended to Andrew a specific one that would benefit both the company and their customers. It happened to be a best practice as well and was known to be used by others operating in the very same industry. Andrew rejected it right away in favor of another option he felt was more appropriate and which happened to closely resemble current business practice. This sub-team was by now very much tiring of Andrew constantly shooting down almost every recommendation they put forth. They were now at a point where they decided to fight back. They strongly disagreed with Andrew and they told him so quite forcefully. Before he had time to tell John about what was now going on within his sub-team, a small group of people from this sub-team, including Susan and Brian, paid a visit to the Controller, Ronald. They explained the options offered by the new ERP software system, the pros and cons of each option, and their specific recommendation. They further told Ronald that their recommended alternative was best for a number of reasons which they further elaborated upon. He listened very carefully as they made their case for their specific recommendation. Then he told them that this all sounded rather fine to him but asked if they had checked in with Andrew regarding this important matter. When Susan said they had and that Andrew had rejected their recommendation, Ronald became very concerned. He told this group to be patient and allow him a little time to discuss the situation with Andrew himself. The group agreed and left Ronald's office convinced they would be vindicated by him.

Ronald caught up with Andrew later on of that same day. He shared with Andrew what had happened earlier. Andrew became enraged telling Ronald that this group deliberately went behind his back after he specifically told them his position on the matter, which he now considered to be closed business. Ronald told Andrew that he tended to agree with the group's recommendation based upon his knowledge and experience in regards to this specific matter. Andrew then told Ronald to stay out of this situation and let him and John take care of it. Ronald agreed to do so knowing full well that Andrew would soon be replacing him and running the show going forward. As you would expect, Andrew paid John a visit to update him on the situation. John of course backed Andrew without hesitation and sought out Brian. He told Brian in no uncertain terms to get Susan to help get the others to back down on this matter and to stop leading the financial and cost accounting sub-team astray. Andrew had made a decision and that was that. The matter required no further discussion. The group needed to move onto their next task or risk jeopardizing the entire project. Brian then talked to Susan to first learn what had transpired and to get her take on things. She told Brian that her sub-team had a lot of pent up frustration with Andrew and John. She told Brian this specific issue seemed to bring this frustration out into the open and that it could no longer be contained.

Brian and George contacted their consulting partner to explain this latest episode directly to him and to seek out his guidance. He really had little to offer them since he was well aware of this client's past behavior, not to mention their lack of making payments to this "Big 6" consulting firm for services rendered to date. So Brian got the members of the financial and cost accounting sub-team together, minus Andrew, and he told them all to yield and let Andrew have his way on this matter and on all others as well. The project needed to move ahead and there was nothing anyone could do to change either Andrew's or John's position. This is not what the group was expecting to hear from Brian, but they certainly were not at all surprised. Susan was extremely quiet during Brian's remarks and she had nothing to say to the group after he left the war room. The matter would soon be put behind them as they trudged along with the balance of their project activities.

It finally came to pass that the project was approaching its termination point. The project team along with the external consultants had endeavored to document the current state, develop the future state, configure and test the features and functions of the new ERP software system that were to be applied within the scope of the project, and get all end users trained and otherwise prepared for the "go-live" date. With the approach of this important final project milestone date, the project team got a bit more engaged in the effort and put aside their many past frustrations. By now, they had grown accustomed to letting John and Andrew (plus the other three company sub-team subject matter experts) get their way. By not resisting, things were far less tense but people remained very unhappy nonetheless. They performed as they were required to do, however, and the project moved ahead as it was intended. The consultants, by acquiescing to John and the other project leaders, finally got paid, albeit after many delays and threats. John and his subject matter experts would go on to claim the endeavor to be a great success and that it was achieved on schedule and within the budget. This was not always the case with other projects that had been undertaken by this company.

The real victim of this project was what was actually achieved versus what could have been accomplished. Given the long and hard efforts expended by so many, not to mention the high cost incurred by the company, the outcome was nothing to celebrate about. With the new direction of the company being dictated by the big changes imposed upon it from the market within this industry, the business processes that were now operating using the features and functions of the new ERP software system were simply mediocre at best. It was not for the lack of trying on the part of both the company project team members and the outside consultants. The results were what they were. John and Andrew would soon go on to earn promotions for jobs well done.

CHAPTER FOUR

Case Three: Project Go-Go

The Challenge

Once there was a large company that set about to implement a state-of-the-art Enterprise Resource Planning (ERP) software system. This company sold a line of products that was second to just a couple of others worldwide in terms of brand name recognition, and they were the leader in their industry by far. They were growing rapidly from not only increased revenues but also global acquisitions. The company decided to undertake a project to implement this new ERP software system because of their rapid growth. By doing this, the company had to find additional sources of capital to finance the initiative. This company had a rather distinctive corporate culture. First of all, it was a mix of "can do" along with a strong "win-win" attitude. So, it was not a surprise to anyone within the company that the project leadership selected a theme for this new project of climbing a mountain — and not just any old mountain, but the world's highest: Mount Everest. The other important feature of their corporate culture had to do with how they treated their employees, which was just like family. If someone's performance was not up to par, instead of putting the individual involved down with negative performance ratings and threats, they rather sought ways to build the person up. If there was some sort of training they could take, then they were instructed to do so and, in turn, were financially supported when they actually did this. Part of the reason for such behavior was the fact that in order to attract good talented people to come work for this company, which was located in a rural area, they had to work much harder and spend more money than many other employers. As a result, they took a good amount of time and expended a lot of effort in selecting the right employees, and then they did everything possible in order to retain them no matter what happened.

With this all said, a senior sales and marketing manager, Harold, was picked by executive management to represent the company as the project manager for this significant endeavor. Once Harold learned he was to be appointed as the project manager, he did some personal research to learn about Project Management and one of the many things he discovered was the value of having a Project Management Office (PMO). Also, because of his recently acquired knowledge about Project Management, he then decided to select two other junior managers to join him in a newly formed PMO to help him run this project. Simon was a business manager and Janet was an information technology manager. In addition, a "Big 6" consulting firm was selected after a thorough due diligence exercise was conducted. This firm put forth a project manager candidate who was subsequently interviewed by Harold, Simon, and Janet collectively. Mark was subsequently deemed to be a good choice and he then suggested that he go out and find two consultants to mirror Simon and Janet's roles within the PMO. They all agreed with Mark and, in due course, Matthew and Linda were brought into the PMO with Mark representing the outside consulting firm. Then, they all began the difficult and time consuming business of determining the other members of the project team required to accomplish the project objectives.

The scope of the project included financial and cost accounting, materials management and inventory control, production planning and control, sales and distribution, and the new business information warehouse module of the new ERP software system, plus all of the technical aspects required by the project. For each of these software modules (including the technical factors), a sub-team was created within the overall project team with a team leader represented by someone from the company along with a corresponding peer from the "Big 6" consulting firm. When everyone was finally determined, the project team numbered one hundred and twenty three people. A very large and open war room was prepared to house everyone, with temporary partitions placed between each of these six sub-teams. Harold was a very firm believer in fair competition due to his sales and marketing background. He knew that these six sub-teams, particularly the five with a business process focus, would likely compete with each other in the performance of the many upcoming key project activities. Harold wanted to take advantage of this energy for the benefit of the overall project, but he was now uncertain as to how best to accomplish this without having any of the downside aspects of such competition. He was determined to find a way to do this in the very near future. In the meantime, he proceeded to ask Mark to identify other clients of this "Big 6" consulting firm, who had recently implemented this very same new ERP software system. Once Mark found a few such companies, Harold reviewed the list and determined that he wanted to have a conference call with the project managers at two of these organizations. He then asked Simon and Janet to prepare a list of relevant and very specific questions to ask these project managers. Harold wanted to learn both what they did right as well as not so right. He asked Mark to make the necessary arrangements. The conference calls were then held in an open and thorough manner. Harold asked his fellow PMO members, including those from the "Big 6" firm, to summarize the results that were obtained from these conference call discussions. Once this was done, the PMO group presented this information to Harold who took detailed notes from what his fellow PMO members conveyed to him. He then asked them for two days to allow him time to analyze his notes and to transfer them into priorities to be applied to their new project.

Once Harold performed his analysis and felt comfortable with what he came up with, he was ready to share his findings with the other members of the PMO. He told them that their first priority would be to come up with a plan of performance rewards for the entire project team including the external consultants. His second priority dealt with

team building and again he charged the others to
seek good ways in order to accomplish this. Finally,
Harold's third priority was about handling issues
and making timely decisions. He asked his fellow
PMO members to take some time to address these
three significant priorities and to come up with a
joint set of recommendations for him to consider. A
week and a half later, this group gathered to share
their thoughts and ideas with Harold. Regarding
his first priority, having to do with project team
performance rewards, they recommended specific
financial rewards be paid out to everyone at key
project milestone points at an increasing value
as the project progressed toward a successful
completion. In terms of Harold's second priority,
having to do with team building, they suggested
a series of specific team building exercises from
several highly reliable sources starting with the
upcoming project kick-off event. As far as Harold's
third and final priority regarding issues and
decisions, they recommended an A-B-C approach
in order to define and categorize project issues to
include specific decision-making turnaround times
along with exactly whom should be directly involved.
An "A" issue would be one of the highest priority,
requiring a twenty-four hour turnaround time and
involving senior executives, while a "C" issue would
be one of the lowest priority requiring a seven day
turnaround time and involving middle management.
Harold was very pleased with all that he heard. The
only thing he further requested of them was that
they add to their recommendations pertaining to his
first priority. He directed his PMO group to provide
for the specific and increasing financial rewards
they suggested at key project milestone points, but

on the condition that such rewards would only be paid out if all the sub-teams met all the deliverable requirements of each given key milestone. Harold remembered about the likelihood of competition amongst the sub-teams, particularly the five business process ones, and he clearly understood that these sub-teams would not have exactly equal amounts of work to handle. Some sub-teams would need to develop more deliverables than would others, while some would encounter more complex matters than would others. So given this situation, Harold sought to foster overall project teamwork while not interfering with sub-team competition up to a point. He then told his PMO group to remind everyone that, despite such obvious and not so obvious differences between the workload of the sub-teams, that they were all collectively responsible for seeing to it that all achieved in total what was required at each given project key milestone. Harold heard something from the project managers of the other companies they previously spoke with that now finally registered with him and this was the solution that he had been seeking earlier.

Aside from these three priorities, the PMO group gleaned a lot of great information pertaining to every other aspect of the project, such as what to include in and how to conduct the project kick-off event through to the requirements for a successful "go-live". Once these ideas were all assembled and fully documented, the PMO team began to plan in detail all the necessary project tasks, due dates, and responsible parties. Further, they worked on developing the slide deck to be used at the upcoming project kick-off event. Once this was all available, they reviewed everything one last time. At the project kick-off meeting, which was a daylong affair, each member of the PMO presented a certain portion of all the material to be shared with the entire project team. Harold purposely did not want to dominate at this most important and first project team event. At the point in the presentation when Janet revealed the performance reward structure for the entire project, Harold interjected some additional clarifying comments. He told the entire project team that successfully completing a given sub-team's work products and activities was just simply not enough. Those members of sub-teams that were finished before the key project milestone point was reached were to then go and help out other sub-teams who were still working hard to finish their own set of deliverables. Unless all sub-teams completed all of the requirements of a given key project milestone, no one would merit the designated performance reward to be dispensed at that point in the project lifecycle. Harold then said that the project plan, including the key milestones, would not change during the entire project lifecycle and that there would be absolutely no exceptions granted for any reason. He explained that a lot of very careful consideration and thought went into the development of the project plan. Yes, the plan was rather aggressive, but it was definitely doable. Harold was quite emphatic about both of these two significant points.

Climbing the Mountain

The project kick-off event came and went and the project finally got underway. The level of excitement was quite high amongst all of the parties associated with this important new effort. First, there was a short period allotted for the "As-Is" to be verified since this company had done a reasonably good job of maintaining its business process documentation beforehand. Then, the project plan called for some time to address new and improved business processes that would take the features and functions of the new ERP software system into account. Matthew had developed the schedule for the "To-Be" workshops and he conveyed the requirements of the template which was to be completed by all of the members of the five business process sub-teams as well as the technical sub-team. He encouraged them all to learn as much about the features and functions of the new ERP software system and associated "best" practices as possible in the time allotted to them for this task in order for them to apply these as appropriate to the future state business processes. The project team relished the opportunity to do so.

The project team's work efforts proceeded without any real hurdles. There was the usual need to re-explain certain matters to specific people, but nothing more than that. The "To-Be" brainstorming sessions were conducted with the utmost of care. It became quite clear to all observers and players that the project team was getting along together rather well, both as an entire team as well as within each of the six sub-teams. The PMO was rather pleased with just how things were going as well, but they knew it was still very early on. It was also becoming obvious to all that the various business process sub-teams had unequal numbers of "To-Be" business processes to contend with although each sub-team had very close to the same headcount. For instance, the financial and cost accounting sub-team needed to develop eighty-five future state business process procedures while the materials management and inventory control sub-team had to do just fifty-two of these. The sales and distribution sub-team had seventy-five to develop and document, but they felt that a large portion of this total were more complex than many of those of the other business process sub-teams. They were all given the same amount of time for this critical project activity. Just as Harold predicted, these five sub-teams began to compete with each other despite clear workload differences. That's why he had reinforced to all the need for teamwork. By way of an example, when the materials management and inventory control sub-team completed all of their assigned tasks for a particular key project milestone, they went right over to aid the sales and distribution sub-team with their continuing efforts. Even though none of them were sales and distribution experts, they could certainly take meeting minutes, make and distribute copies, answer phones, etc. In this manner, all of the sub-teams were able to fully complete all of their specific tasks for this key project milestone at the designated time. As a result of this performance, the first round of financial rewards were paid out as was agreed to at the beginning of this project and communicated to all project team members at the initial project kick-off event. This of course made everyone participating in this effort quite pleased.

At each project milestone there was some time designated for team building exercises. The PMO had pre-determined the specific materials to be utilized for this rather important ongoing project activity. Each exercise selected was done to achieve a certain aim or make a specific point pertaining directly to this project. Further, the exercises were also picked with the objective of allowing all of the many members of the project team to both have some fun while getting to better know their fellow team members in a social environment instead of a stressful project work atmosphere. With such a large project team, it wasn't easy nor did time usually allow for getting to know people on the other sub-teams, except during such periods. Everyone said they got a lot out of these experiences, including having a good bit of fun doing so as well. These periods went a long way towards relieving any built up stress from performing work activities. These exercises were not very expensive to execute, nor did they require any separate or special facilities in which they needed to be performed.

As the project moved into the configuring and testing lifecycle phase having to do with both the new ERP software system, everyone was now feeling even more comfortable working together and in performing all of their required work tasks. The PMO members now had a high degree of confidence in the project team members. This was based upon their constant review of the many work product deliverables generated to-date. The documents had a great deal of detail, were well written and clear, contained facts and figures to support the recommendations put forth, and adhered to all of the specific requirements of the various templates provided by the external consultants. This opinion was further validated by the ease and timeliness

in obtaining business process owner approvals.
In addition, the company project team members
regularly reported that all of the outside consultants
were doing their very best at continually transfer
knowledge about the various configuration options
and the consequences of each alternative. In this
manner, the company project team members could
weigh the pros and cons of each option before finally
arriving at a final decision which they felt was
the very best fit for the particular circumstances.
Finally, issues and decisions were handled
according to the A-B-C approach. Issues and their
corresponding resolutions were carefully documented
and tracked while decisions were handled using the
approved and pre-determined level of management
authority required based upon the significance of
the specific matter at hand. Everyone was fully
complying with all aspects of the project charter.
The project scope was challenged several times and
each time the required efforts were handled per
the approved and pre-defined change management
process. In the end no significant scope changes
were permitted.

It Just Takes One

There is no such thing as a flawless project for if
that were the case there would be no need for a
project manager, yet alone a PMO. Within the
financial and cost accounting sub-team there was
a fairly new employee named Paul, who had been
the company Treasurer prior to joining this project
team and he had a bad attitude. This stemmed
from two principal factors. The first was that he
had been a part of the selection committee when

this new ERP software system was initially selected. Paul did not support the overwhelming decision by this committee to choose this particular ERP package. He felt it was too complex and far too costly for the company. Paul had only been with this company for just about six months when he got tapped to participate in this important activity. As a result of both his involvement in this committee and his expertise in a particular aspect of financial accounting, he was ultimately selected to join the project team for the implementation of the new ERP system. The other reason for Paul's poor attitude was that by joining the project team in the war room, he had to give up his private corner office and administrative assistant who brought him his coffee and Wall Street Journal each and every morning. Instead, he now had, just like everyone else on the project team, a small cubicle with just enough space for his laptop, plus an additional shelf above for storing documents and manuals pertaining to the new ERP system and his own project work papers.

Because of this, Paul began to act out on his frustrations typically by yelling at his computer screen, pushing his chair back from his cubicle, and then leaving the area for parts unknown for extended periods of time. This behavior was most disruptive to his fellow financial and cost accounting sub-team members seated near him. On his left, sat Rachel, one of the financial and cost accounting sub-team's external consultants. To Paul's right sat Cynthia, a company employee who was this sub-team's subject matter expert regarding all accounts receivable business processes. She had ten years of experience with this company starting out as an accounts receivable clerk and rising to become

the accounts receivable supervisor through a lot of
hard work and a bit of good fortune. Unlike Paul,
she had no private office prior to joining the project,
nor an administrative assistant, plus she was not
a company manager, which meant that she was
not yet a salaried employee. But now, in terms of
their respective roles on the project team, they were
treated by all parties as equals. This mattered very
little to Cynthia but, unfortunately, it mattered
very much to Paul. Come five o'clock each work
day evening, Paul could be found packing his things
and preparing to head for home while Cynthia was
usually still busy going about her various project
work tasks. They both were parents with spouses
and children at home requiring their full attention.
Somehow, Cynthia found ways to accomplish both
of her responsibilities rather well while Paul did
not. Bradley was the financial and cost accounting
sub-team leader representing the company and he
was supported by Adam from the "Big 6" consulting
firm. Adam became troubled by Paul's disruptive,
unprofessional, and rather poor behavior. He
brought this subject up with Bradley who agreed
Paul was becoming a problem. They decided to
speak to Paul together about it, which they did.
Paul reacted rather surprised and told them both he
would cease acting out in this manner. Both Bradley
and Adam came away from this brief encounter
feeling the matter was now over and done with. But
in less than a week, there was another unsettling
flair up while Bradley and Adam were in another
meeting away from their war room office space. Paul
upset Cynthia, who in turn shared the incident with
Rachel, who had witnessed some of what had taken
place. They decided to report the matter to Bradley

and Adam upon their return later that same day.
Upon learning about this situation, they asked about
Paul's whereabouts, of which no one seemed to know
for sure. It appeared to all of them that he had left
work for the day and had simply gone home early,
without notifying anyone.

The next morning, Bradley and Adam pulled
Paul aside into a small conference room to question
him regarding the events of the day before. Paul
told them he was very disturbed by certain features
and functions of the new ERP software system that
were extremely complex and in his opinion much
worse for the company than current business process
practices were in terms of his area of expertise.
Bradley spoke first, saying to Paul that even if what
he indicated was all true, it in no way justified his
bad attitude and poor behavior. Paul then said he
felt bad about what had happened and that he would
definitely apologize to Cynthia for upsetting her.
Bradley and Adam were now concerned about Paul,
but they still kept hoping it would all go away given
Paul's sincere confession and intended apology to
Cynthia. Unfortunately in life, both hope and good
intentions do not make real problems disappear.
Paul actually never apologized to Cynthia and just
several days later had a repeat performance of his
earlier meltdown, only this time he disappeared but
did not go home. Instead, he returned to his former
office area where he spouted off to many others in
the vicinity about his total lack of confidence in
and support for the new ERP system now being
implemented. His listeners became rather disturbed
about his comments and word of this event soon
got back to Bradley. Bradley decided to focus his
attention upon other more pressing project activities

and put aside this matter involving Paul at least for the time being. Rachel, in her travels amongst many financial and cost accounting end users, picked up on their concerns which arose as a direct result of Paul's recent remarks putting down the new system. She then told Adam about this discovery and he became very worried about Paul and now even further about Bradley's lack of reaction to Paul's continuing and increasing bad behavior. He too knew that there were many other project work priorities that demanded his and Bradley's attention right then and the very last thing they both needed was to spend any more time on Paul's outbursts. But Adam now realized from his prior project experiences that much more had to be done to address the situation concerning Paul's poor attitude and disruptive behavior. The fact that Bradley was refusing to deal with the matter only made Adam more convinced he had to try something else. The first place he went was to visit Mark to provide a history of the events to-date in regards to Paul's acting out. Now Mark was not totally unaware of the situation since he, along with some of his fellow PMO members, had indirectly heard about some incidents involving Paul and Cynthia.

Mark listened closely to Adam and then suggested they both go pay Harold a visit to discuss the situation concerning Paul. Mark had Adam explain the events to-date resulting from Paul's various disruptions. Adam also conveyed to Harold the several conversations he and Bradley had with Paul only to witness the same behaviors reoccur within days of their discussions with him. Then Adam shared his more serious concern about Bradley's apparent lack of interest and the fallout from Paul's comments to several financial and cost accounting end users. Harold took it all in while asking several pertinent questions in order to help clarify his understanding of what he was hearing from Adam. He then asked Mark and Adam to leave so he could consider what next steps should be taken. Two days later, he summoned for Mark and Adam to come back to his PMO office. He first asked Adam if anything new had transpired since their last encounter regarding the subjects of Paul and Bradley. Adam replied that nothing new had taken place since their last conversation. Then, Harold proceeded to explain to them both something he knew neither of them was aware of. He said that prior to the time when each company project team member was brought onto the project, they each were clearly informed of two very specific things. The first of these was that they would not be allowed under any circumstances to return to their former jobs within the company until the project was completely finished. The second thing they were told was that they would do everything in their power to make this effort a success for the company. Then, Harold told Mark and Adam that each person was asked to sign a written contract which contained these two explicit statements that he just shared with the two of them. The point for revealing this news to Mark and Adam was to inform them both that despite Paul's inappropriate actions, he was not able to release Paul from his contractual project commitment. Harold further explained the rationale behind this all. He them told them both that Paul's behavior clearly demonstrated to him that Paul was acting out in the distinct hope that he would be let go from the project in order to return to his former role within the company. The project policy established just prior to the beginning of this project of not letting people leave the project was purposely put in place to prevent disgruntled people like Paul from returning to their former jobs in order to negatively mouth off about the project to try and sabotage it. In fact, Harold indicated that he had done his own investigation behind the scenes and he was now convinced more than before that Paul was behaving in such a manner knowing that since he could not get released from his project agreement he would do whatever he could to seriously damage the project in the eyes of as many end users as possible and, by so doing, hope that they would spread negative comments about the new ERP software system as well. So the bottom line was that Harold would not allow Paul to be released from the project. After a few minutes of consideration, Mark and Adam suggested to Harold that he might want to entertain the idea of firing Paul from the company, which would have the added benefit of removing him from the project team. Harold's response was quick and to the point. First, he told them that he had also spoken with the Vice President of Human Resources about Paul and together they decided this was not an appropriate action to be taken at this time. He then further explained that while it certainly was not an official documented company policy, their corporate culture was such that it would not allow for Paul to be fired. Instead, he and Bradley would now be tasked with working even harder to rehabilitate Paul in order to change his poor attitude and bad behavior. None of this was what Mark and Adam had expected to hear from Harold. They said that it was not their intention to get Paul fired from the company; rather, all they wanted to do for the sake of the project was to simply get him removed from it. Now, obviously this was not going to happen according to Harold. Even more importantly, they really understood why and agreed with the policy that had been established and with the rationale behind it. In fact, they were both quite impressed that Harold and his fellow company PMO members had instituted it to begin with and actually stuck to it when it counted the most. So they proceeded to thank Harold for his frankness and time, and then quietly left his office.

Bradley and Adam continued to work on Paul's recovery program while Paul pursued his poor behavior, though less frequently and with a bit less energy. Harold and Bradley had read Paul the riot act in order to make a serious impression upon him. It obviously had an impact, but was not a total cure all, which is what they unfortunately expected. Cynthia and Rachel were relieved that Paul was acting a bit more controlled and less disruptive, but he remained rather frustrated, uncaring, and most uninterested in the project other than getting it over and done with as soon as possible. Paul wanted to return to his former position and office space within the company. At least Paul performed what was asked of him reasonably well and for the most part kept out of any serious trouble with Harold and Bradley. He even had moments of sincere and rather good contributions to put forth but by now everyone he encountered on the project team was just a bit suspicious about his intentions. Whatever damage to the project he had tried to cause was forgotten by the end users he had tried to negatively influence because so many other much more positive

aspects were now being conveyed from a good many
other more reliable sources.

Reaching the Summit

The project methodically progressed forward from
one key milestone to the next. On just one occasion,
a couple of sub-teams came up short in terms of
their work deliverables resulting in missed financial
rewards for the entire project team. Despite this
blip, the project moved along quite well delivering
very positive and rather high quality results.
Overall, the project team was performing very well
for such a large and rather diverse group of people
working on such a new and complex information
technology endeavor. The project plan for the most
part was adhered to and was never once changed in
any significant way. At times people would complain
about the aggressive timetable, but then they just
got down to work and tended to produce quite well.
There were a few surprises along the way, but
nothing that couldn't be handled by the PMO and
the project team, and definitely nothing that caused
any real delay or work stoppage.

Harold was now beginning to see the fast
approaching goal line of this project in plain sight
and he was feeling really good about how things
were progressing. The situation regarding Paul
was contained, but certainly neither forgotten nor
ignored all together. Small and rather short flair
ups occurred which were put down forcefully and
quickly by Bradley and Adam. Paul's issues did not
seem to prevent the financial and cost accounting
sub-team from delivering well on its project
commitments. The PMO group as a whole was quite

83

impressed by the high level of quality produced in the many project team deliverables, particularly in how often they took advantage of best practices and the many features and functions offered up to them by the new ERP system. This view was echoed by the business process owners and senior technical managers as well. In addition, the PMO was pleased by the high degree of teamwork and sub-team cooperation exhibited by all. This is one of the most important factors in an ERP implementation project because of the extent of system integration that is involved. And this specific ERP software system had a reputation in the marketplace as being both complex and rather expensive. Both criticisms were accurate, but what was too often overlooked was the tremendous flexibility the system could provide to end users and to a large global company as a whole. There surely was a price to be paid in order to achieve this flexibility. Once it was operational for a time, end users could enjoy a very powerful information management tool for not just processing their transactions efficiently but also for very effective reporting and analysis purposes. Such information was invaluable to a company in order to gain and maintain a strategic competitive advantage in the marketplace.

And so, by being both on schedule and within budget, this project approached its "go-live" date rather well prepared for the big cutover from its disjointed legacy systems to just one primary management information system based on the new ERP software. Everyone was now looking forward to closing out this effort with rather mixed feelings. They were glad the project was fast approaching its conclusion and were very proud of what they each and collectively had accomplished, not to mention what they had learned from the entire experience. At the same time, however, they were a bit sad about having to say good-bye to so many new friends, including those from the "Big 6" consulting firm.

This project was ultimately declared a success by the executive management at this company and a good many of the perceived benefits soon came to fruition. Harold and his fellow company PMO members, Simon and Janet were soon promoted within the company, and rightly so. All of the other company project team members, including Paul, returned to their former positions now as super-users of the new ERP software system. Getting to use what they helped to create and now managing others in its daily usage was their ultimate reward for all of their very hard work and long hours that proceeded "go-live". Also, working to bring newly acquired global businesses onto this ERP software system was another important activity which reminded many of them of their past days on the ERP software implementation project.

Some nine months after the "go-live" date, a post audit of this project was undertaken by the newly formed permanent PMO group. No one from the original project team was involved in this additional effort. During this audit, it was discovered via a survey that end users were very pleased in using the new ERP system's features and functions. They were rather comfortable in their understanding of it, and in their ability to perform all of their required business process activities. They were already considering requesting system enhancements which not long thereafter resulted in an add-on project to implement certain specific additional features and functions of this ERP system. Once the post audit was concluded and the results released, they clearly showed that in almost every category analyzed this project had been a very big success for this company. In the all important ROI ratio, the result achieved exceeded expectations by a fairly wide margin. The company expected this outstanding performance to continue for some time into the future.

CHAPTER FIVE

Lessons Learned

Each of the three cases that were just presented represents one of the three distinct forms of information technology project outcomes typically found:

- Utter and outright project failure

- A reasonably fair project result

- A rather good project outcome

By analyzing these three types of project results, specifically in terms of the three plays just presented, we can get an even better perspective and understanding about project critical success factors which tend to make a real difference in the final outcome. First, from the vantage point of what was in common across all three of these project plays we find that:

- They all seemed to have the proper number of good people involved in these efforts

- They all appeared to initially have the proper level of dedication and effort put forth

- They all seemed to initially have the proper amount of commitment by all of the parties involved, including the external consultants and executive company management

- They all had adequately qualified project managers

- They all seemed to have a reasonably good and sufficiently detailed project plan for the effort that was subsequently undertaken

- They all appeared to have a clearly defined project scope, which for the most part, they held to throughout the entire endeavor

- They all had a reasonable budget for the effort that was undertaken

- They all had a reasonable, but aggressive schedule for the effort that was undertaken

- They all employed good project management tools in planning and controlling their project activities

- They all employed outside consultants with the relevant skills and experience for the effort that was undertaken by first recognizing that they even needed such support and allotting the funds to pay for this service

- They all had rather clear and well intentioned business purposes and goals established going into these projects

The following items describe how each scenario shows how critical project activities were handled differently, including the results that were achieved:

- Project No-Go

  - No regard was ever given to any team building throughout the entire endeavor resulting in a total lack of project team unity and cohesiveness

  - The good knowledge and advice of the external consultants was never considered, nor ever really seriously taken into account

  - The company project managers were detested by all of the project team members due to their personal bad behavior and rather poor attitudes

  - The project managers regularly tried very hard to run the project as a dictatorship— or worse—as a game to be won by and for themselves alone

  - No project team member incentives were considered, provided for, and none were ever dispensed by the company

  - Project team member issues were treated as if they represented an anarchy by the project managers in order to maintain their own authority and sense of total project control

  - End users never got to use any of the new ERP system's features and functions. No "best" practices were ever employed. No business benefits were ever derived by the company or by any of its customers.

- The project was terminated prior to its "go-live" date due to a lack of available company project team members. It simply crashed and burned prior to its completion.

- The project was running somewhat behind schedule and was a bit over budget when it was terminated by executive management. As a result, all of the money spent on this entire failed endeavor was a total loss of company funds.

- Project So-So

  - Very little regard was given to any team building initially and throughout the entire endeavor

  - The good knowledge and advice of the outside consultants was never considered, nor ever really taken into account

  - The company project manager was detested by all of the project team members

  - The company project manager totally ran the project as a dictatorship, he constantly acted as all knowing, and he constantly acted as if he was and needed to be right all of the time

- No project team member incentives were considered, provided for, nor were any ever dispensed by the company

- Project team member issues were treated rather brutally by the company project manager in order to maintain his sole authority and sense of absolute control

- End users ultimately had rather limited usage of the new ERP system's features and functions. Few if any best practices were employed. Some minor business process efficiency, reporting and analysis, and information technology improvements resulted from the use of the new ERP system's features and functions. There were a lot of unfulfilled expectations by both internal users, and even more importantly, the new and growing commercial customer base.

- The project came in on schedule and was within budget at the "go-live" date. The ROI result was a good deal below of what could have been achieved.

- There was no post-audit ever undertaken regarding this project to gather lessons learned for any future efforts

- Project Go-Go

  - A great deal of consideration and regard was given to team building prior to and throughout the entire endeavor

- The rather good advice of the external consultants was highly valued and always considered. This relationship resulted into a real solid business partnership.

- Adequate preparation time and effort was invested during project planning, including seeking out the experiences of other outside parties who preceded them with a similar ERP system implementation effort. In so doing, they learned firsthand what worked rather well and what did not.

- The company project managers were highly regarded and respected, plus a Project Management Office (PMO) approach was fully utilized throughout this effort and was carried forward as a permanent organizational unit

- The company project manager never attempted to do it all, nor did he ever try to be all knowing and always right

- Project team member incentives were provided for and were dispensed at each key milestone point as long as the specific goals called for were fully achieved by all

- Project team member disagreements were always treated fairly by the company project manager and others' views were certainly always taken into serious consideration. In addition, prior to the start of the project, anticipating how to handle issues was carefully planned for by the project manager and PMO.

- End users were very pleased and quite comfortable in using the new ERP system's features and functions. A great many best practices were employed and before too long, an add-on project was undertaken to increase the company's usage of the ERP system's vast store of features and functions.

- The project came in on schedule and within budget as was planned

- A post audit was conducted nine months after the "go-live" date to assess project success or the lack of it. The results clearly indicated that the endeavor was a large success from a good number of viewpoints, including achieving better than expected ROI.

In conclusion, just from these three project plays alone, it should be clearly evident that people challenges matter greatly and can have a negative impact upon a project, overshadowing many other positive factors. Good project management discipline can help contain—or even better yet, eliminate—people issues if they are applied early before any real conflict gets a chance to arise and take hold. Further, we can see the influence of organizational culture and internal politics as having a definite impact in contributing to people challenges in project situations. This was certainly the case in Project So-So where the macho-military culture clearly had a great deal to do with how things played out. On the other hand, the organizational culture in the Project Go-Go play significantly contributed to the success of this project situation. Organizational culture is defined as the collective behavior of people sharing a common corporate vision, goals, and values. This further includes shared beliefs, habits, working language, systems, and symbols. Internal politics has to do with the fact that people tend to like to be part of a group. Prior to the time when the concept of business processes was as widespread as it is today, work was typically performed by people working in functional groups (i.e. silos). Today, business processes go across many functional areas creating new groups of people who share a common purpose. Regardless, groups tend to have rivalries which too often can turn into people conflicts. This is also true of sub-groups within a large project team. Both organizational culture and internal politics can definitely influence people's behavior in the workplace, and not always for the better as we have just witnessed in the three cases discussed earlier.

Part Two:

Project Leadership

CHAPTER SIX

What is Project Leadership?

The global profession of project management is currently one of the fastest growing professions today. According to a study commissioned by PMI, as part of the development of their own 2007 strategic plan, project-related activities accounted for 25% of all the money spent by all organizations worldwide. KPMG performed this study during 2003 and once again in 2005, contrasting the results to reveal the following rather surprising statistics:

- The number of new projects increased by 81%
- The level of project complexity increased by 88%
- Project budgets increased by 79%

Now given that these very significant percentage increases occurred in such a relatively short period of time, even more so when you consider that this time period was just shortly after Y2K (Year 2000) and during the time of the Dot-com meltdown. These findings alone should generate rather serious concern by the project management profession, only adding to the many challenges of doing projects well, especially given the rather low success rates for projects that were mentioned earlier. When you factor into these findings the fact that on average 80% of all project issues are directly attributable to people, and that 70% of a project manager's precious time is spent on non-value added activities, I believe you truly have a recipe for a significant professional challenge on a global scale with an increased likelihood for project failure, financial loss, and disappointment. However, the profession continues to place more of its emphasis upon both the process and technical aspects of project management instead of the much larger challenge having to do with people issues. This emphasis is borne out by all of the evidence I found when reviewing the nature of the profession's research, the seminar agendas being issued, the books being published, and the subject matter being taught and tested in the PMP preparation classes and certification exam. I have come away with the distinct impression that it is far less challenging and apparently more appealing to develop new and improved tools and techniques, or work with new forms of technology, rather than to find real and lasting ways of improving human relationships in the performance of project management activities.

Sometimes, when describing a term or concept, it is easier to begin by discussing what it is not before turning to trying to define what it is. This is exactly the approach I will now take to explain project leadership. The first Standish Group CHAOS study, dated from 1994, revealed the following set of information technology project outcome results:

- 31% of all projects were cancelled prior to their completion (e.g. Project No-Go)

- 53% came in at a cost that was 189% greater than their original cost estimates while taking 230% longer than their original time estimates

- Only 16% were completed both on time and within their budget, but with only 38% of their original user feature and functional requirements met

Ten years later, in 2004, this very same study was conducted yet again by The Standish Group. This time the project outcome results revealed that 34% of all projects were completed successfully. This more than doubled the project success results, but was still far below a 50% break even rate. As part of the findings, the study discovered that many more small projects were being undertaken and project management best practices were employed to a much greater extent. Despite all of the many new techniques and tools in the global project management marketplace, only one third of all information technology projects were deemed to provide an overall successful outcome. In this study, success was measured by the project management profession's typical key performance standards, which are being delivered on schedule, within budget, and even more importantly, by satisfying all customer (both internal and external) requirements.

The three challenges of project management need to be explained just a bit further. Process issues tend to be associated with such important project matters as determining the particular methodology, approach, and the proper tools to employ in the performance of all project activities. Technical issues involving information management system projects have to do with determining the computer hardware, networking equipment, security aspects, and software solutions necessary to accomplish the entire project effort. People issues pertain to just about everything else involved with a project. Further, it now needs to be made quite clear that most project challenges occur within more than one of these three categories. It is far less likely that only one category completely describes a specific situation. Despite this being the case, people issues tend to dominate above all else, even when another category also applies. For, when people are in a state of conflict, project issues take on a whole different level of challenge and effort in order to get them resolved. While I was studying in preparation for taking the Project Management Professional (PMP) certification examination, I learned that there are seven sources of conflict involved in the practice of project management. These are listed below in priority sequence as follows:

- **Schedules** – This represents both a people and process issue. Often experts will disagree about the duration needed to complete specific project tasks.

- **Project priorities** – This also represents a people and process issue. Priorities may change during the project lifecycle and could well be influenced by external business events (e.g. a business acquisition or divestiture, significant organizational restructuring, etc.).

- **Resources** – This also represents a people and process issue. This is typically a sensitive and rather potentially contentious aspect in the practice of project management. Not all Resources pertain to people, but a good many do.

- **Technical opinions** – This represents both a people and technical issue. Experts often disagree on specific technical matters.

- **Administrative procedures** – This represents a people and process issue. There are clearly multiple ways to accomplish specific project activities.

- **Costs** – This also represents a people and process issue. This too is a rather sensitive aspect in the practice of project management requiring both accurate and complete cost information.

- **Personality** – This final source of conflict is purely a people issue. What is most significant and rather surprising is that this source of conflict is seventh out of seven in priority.

The point here is that these represent only potential sources of conflict. Once the spark is lit, however, they can rapidly grow out of control. If a project manager works with all of the people involved by effectively and thoroughly discussing these important matters and then coming to some reasonable resolution, they are simply nothing more than initial points of disagreement amongst fellow experts. It is my contention, that if people working together on a project are in a state of conflict, then any project issue becomes a people challenge first and foremost. By the same token, if the conflict between people engaged in project work activities encounters a clear cut process or technological issue, the matter can more easily be resolved just so long as there is an absence of any kind of real people challenge. Given proper attention, people with the necessary knowledge, experience, and adequate time and money, will successfully resolve a process or technical issue, which will enable the project to proceed. However, people issues that cannot be contained nor resolved are deadly to a project's chances for any kind of successful outcome. Disputes over important matters such as schedules, priorities, resources, etc. are solvable only if there are no serious people issues threatening; otherwise, all bets are off.

Surprisingly few reasons that people issues tend to develop have to do with such obvious differences in people's sex, age, race, culture, language, and religion. One aspect that does aggravate people challenges is what is known as the "Generation Wars." In today's workplace, there co-exist three distinct generations. These three are known as the "Baby Boomers" (people who were born between

1943 through 1960), "Generation X" (people who were born between 1961 and 1981), and "Generation Y" (people who were born in 1982 and thereafter). There are different values and priorities held by each of these generational groups that can be a source for people issues arising during project work activities. These issues could turn into real conflict unless they are proactively prevented. In addition, it has been noted by researchers that we as project managers are working under certain bad assumptions. The first of these is that we assume that adults in the workplace know how to work well together and that they will automatically just do so. Secondly, sometimes we force everyone to behave in the very same manner, while not recognizing individual versus collaborative efforts. Finally, we sincerely believe that all problems can be resolved by simply discussing them with each other in a logical manner. people brought together to work as a team in order to accomplish a series of critical project tasks within the typically stressful schedule, cost, resource, and quality constraints of an information technology project often have had no such prior project, nor team experience. Even if they have had such a project experience, it was likely not a very good one based upon the project failure rate statistics already mentioned. We cannot simply expect people to work together well as a cohesive whole without team-building efforts. Forcing everyone to act in the same manner is simply unrealistic. Much like the current state of the United Nations, expecting people to recognize and then solve difficult problems by discussing them face-to-face has often proved to be rather naïve, most frustrating, and has provided a rather false sense of comfort. The real proof is in the results obtained, and in projects clearly demonstrating that this technique alone does not work well in the presence of significant people issues. For, if it did, the success rate for projects would be significantly better.

What tends to make people issues come out and rise to the forefront during projects are:

- The uncertainty of the project tasks – "We have never done this before."

- Resource increases/decreases can result in overload/boredom cycles for individuals, frequently leading to infighting.

- One problem can generate or influence yet another problem.

- Project managers have rather limited formal authority and span of control.

- Volatile situations develop as a rule, rather than as the exception – primarily due to stress.

- Authority of position usually has less of an effect on project success than the ability to influence people in other ways, such as with one's technical knowledge, and thus things may seem out of control.

Another set of factors contributing to people issues is the very nature of the types of people involved with project oriented work. The people required are typically very technical in terms of their personal backgrounds. Surveys conducted have revealed that such people tend to be as follows:

- Highly detail oriented

- Too often reluctant to share information

- Less concerned about users who simply seek quick practical applications

- Are rather good starters of things, but not finishers

- Demand a lot of freedom and flexibility in the performance of their jobs

- Are rather high achievers

- Tend to be anti-structural (they absolutely detest bureaucracy)

The overall conclusion to be drawn from these survey results is that technical people, generally speaking, do not possess very good people skills. This is clearly contributing to why people issues tend to arise in project situations, where a predominance of technical people is a rather necessary component. The education and training that technical people receive is primarily focused upon analyzing and solving technical challenges. Sorely lacking is soft skills awareness and training, the most obvious of which has to do with interpersonal communications. It has been said that in the Information Age, we have created many more ways of communicating with each other than ever before in the entire history of mankind and yet, unfortunately, our effectiveness when it comes to performing good communications is still no better than it was before.

The antidote to people challenges is project leadership. It is rather different from project management. First of all, leadership is defined as "an ability by a person to influence other people". Bear in mind that this influence can be for good or bad intentions. Leadership implies that someone is leading other people, who are willing to follow or be lead. Often, leadership comes not from holding a position of authority, but rather from an ability to inspire other people to go on and accomplish many great achievements, taking on new and very uncertain challenges in the process. While a great deal of written information exists about what leadership is all about, for a great many people it is far easier to describe leadership than it is to precisely define it. Most people have a mental image of leaders whom they have encountered in their lives and can share specific attributes that these people demonstrated:

- They are willing to guide and serve other people

- They hold some sort of a technical competency that they are most proficient at

- They seek, as well as take on, responsibility

- They have an ability to make sound and timely judgments or decisions

- They purposely set themselves out as an example for others to follow

- They know their people, and continually look out for their general well being

- They keep their people regularly very well informed by communicating to them constantly in a clear and consistent manner

- They have a clear bias toward actions and results

- They are optimistic

- They have a clear sense of purpose

- They have great self knowledge and awareness

- They have an ability to be able "to walk in someone else's shoes," demonstrating empathy

- They regularly display real passion, confidence, determination, and persistence in all they do

- They have a vision of the future and seek to turn it into reality

Another way to explain leadership is to contrast it with management. As shown in Table One below, there is a very clear distinction between these two concepts and practices:

Project Management	Project Leadership
Task oriented	People oriented
Administer	Innovate
Ask how and when questions	Ask what and why questions
Focus on systems	Focus on people
Do things correctly	Do the right things
Maintain	Develop
Rely upon control	Inspire trust
Have a short-term view	Have a long-term view

Project Management	Project Leadership
Accept the status quo	Challenge the status quo
Imitate	Originate

Table One – Project Management versus Project Leadership

How this relates to the practice of project management and having to regularly deal with people issues follows. But first, the Project Management Institute (PMI) defines a project as having the following three very specific characteristics:

- It is temporary by its very nature – with a definite beginning and a clear end point; thus, there is a specific time duration

- It is unique in its function

- It is "progressively elaborated" (i.e. developed) over time

Given that project work is thus deemed to be very different from other forms of work, the approach to be taken in handling projects must be performed rather differently as well. This means having people with the relevant skills and knowledge, experience, willingness, and proper good attitudes to take on such an endeavor. I am not just referring to the project manager, but to all of the members of the project team. And I further contend that project work is, simply stated, not for everyone. PMI provides a project management framework that contains five process groups: Initiating, Planning, Executing, Monitoring and Controlling, and Closing.

By applying the concept of Project Leadership across these five process groups, I have developed a reference guide for project managers to use as follows in Table Two:

Process Group	Project Leadership Characteristics
Initiating	Learn to accept people and their needs. Exercise wisdom. Be courageous about challenging and developing people. Help people accept their own project responsibilities.
Planning	Learn to accept the project sponsor's conflicting needs. Help the project sponsor prioritize needs. Learn to accept that project team members will develop a good portion of the project plan details. Learn to accept that others can make better judgments without giving up your own project management responsibilities. Learn to accept the diversity of people's viewpoints.

Process Group	Project Leadership Characteristics
Executing	Learn to accept that others will control certain project work activities.
	Learn to accept project changes as long as they are handled within a disciplined control process.
	Learn to accept that the organization has other important work to accomplish besides this project.
	Learn how to facilitate people without dictating to them.
Monitoring and Controlling	Learn to accept that project work is hard and stressful on people.
	Have the courage to uncover project performance issues.
	Have the courage to reward and recognize good project performance along with appropriately acting upon any poor project performance.
	Promote project member teamwork throughout the endeavor.
Closing	Be courageous about collecting and analyzing "lessons learned".
	Help people secure other appropriate work opportunities after the project is fully completed.

Table Two – Project Leadership Across the Five Project Management Process Groups

Even though a project manager will not perform all required project activities alone, he or she has responsibility for mitigating and proactively resolving any and all people issues that arise — and arise they most certainly will. In order to be able to do this job efficiently and effectively, the project manager must have project leadership abilities in addition to all other relevant and required technical and business skills. Project managers face circumstances not faced by leaders of other types of work endeavors. They must provide leadership in a cross-functional and cross-organizational or matrix environment, which is quite often today a global working environment, deal with conflicting stakeholders, and operate without any formal authority over all of the project team members who are assigned to their project, including external consultants and virtual participants. I believe that the important job of being a project manager is one of the most challenging assignments one can take on.

CHAPTER SEVEN

How is it Best Employed?

N ow that we understand the importance of project leadership, as distinguished from that of project management, we can further examine how best to employ project leadership best practices.

First, researchers say that project managers tend to exhibit seven different leadership styles which are as follows:

- Directing or telling other people what to do and how to do it

- Facilitating or coordinating the inputs of other people

- Coaching or instructing other people

- Supporting or providing assistance to other people along the way

- Being autocratic, or making decisions without input from any other people

- Consulting or inviting the ideas of other people

- Performing problem solving in a group setting, with any decision making being based upon the group's overall agreement (consensus)

There is no single leadership style that is deemed to be the best, nor does a project manager just operate with only one style throughout the entire duration of a given project. A good project manager knows himself rather well, and can apply these seven different leadership styles to fit specific circumstances as they arise during the project's entire lifecycle.

Second, developing one's project team into a high performance team is critically significant. There are five processes or stages that a project team typically goes through in order to achieve this level of status which is as follows:

- **Forming** – This is the initial period of coming together as a project team. During this most important time, the project manager needs to provide clear direction to all of the project team members involved.

- **Storming** – As people on the project team begin to feel each other out, typically during the Project Initiating and Planning stages in a group, people issues resulting in conflict will often reach their pinnacle. The project manager needs to proactively manage any conflicts and work hard to foster win-win relationships and positive outcomes.

- **Norming** – As people issues begin to settle down, during execution of activities in the

Project Planning and Executing process groups, people tend to proactively avoid getting into any further conflicts with other project team members as they begin to seriously focus upon the project's deliverables and objectives, and most especially, their very own work project activities. The project manager needs to provide supportive feedback, while maintaining a clear focus upon the project's objectives.

- **Performing** – Typically, during the Project Executing and Monitoring and Controlling process groups, people focus much more on getting the job at hand done well and doing it on time, which is why this is often the most productive period. The project manager needs to encourage people to strive for continuous improvement in everything that they do.

- **Adjourning** – This, by definition, happens during the Project Closing or the final project lifecycle stage, as the project prepares to be shutdown. The project manager needs to see that all work activities are handled well and in a rather controlled manner.

Project leadership encompasses developing people, communicating with people, handling a great deal of personal stress, solving problems creatively, and managing time, which is often the most precious project resource. Furthermore, information technology projects usually bring along with them many additional hurdles having to do with new and challenging technologies, management's business

115

process re-engineering goals, all organizational change management repercussions, and the required amounts of education and training. Every project team member must strive to achieve project leadership, not just the project manager. In so doing, not only are they aiding the project manager but also they are giving the project a much better chance for success. People may sometime in the future desire to become a project team leader, or even a project manager. Understanding and then applying project leadership skills early on will go a long way towards making this journey possible and much more likely to be successful.

Leading people working together in a project team environment requires some additional comments. The project manager, as always, is responsible for the achievements, or lack thereof, from the project team. As such, identifying the early signs of what is known as a "dysfunctional" team is a necessity. You cannot strive to solve a problem until you first recognize that there is one to begin with. By addressing the following four questions, the project manager can better assess whether his or her project team is dysfunctional or is not:

- Are the project team members openly communicating their views and opinions?

- Does each and every project team member contribute to the project's key objectives?

- Do the members of the project team actually enjoy, actively participate in, and really feel like their project meetings are quite productive?

- Do the members of the project team place a higher value upon the results of the project versus their own personal interests?

By carefully examining the honest answers to these four key questions, a project manager can make a proper assessment about the project team. If they are operating in a dysfunctional manner, which really should not come as any kind of surprise to the project manager, then the very next step is to get to work on turning this bad group behavior around for both the betterment of the project. The goal of any project team should be to become a high performance team along that accomplishes the key objectives of the endeavor. Moving from a dysfunctional to a high performance team takes much time and effort, mainly on the part of a project manager. A dysfunctional team typically results from two root causes:a lack of trust and poor communications amongst its members. Both of these two root causes must be openly dealt with if there is to be any chance for an overall performance improvement. According to Patrick Lencioni, there are five behaviors that a dysfunctional team tends to display, which are as follows:

- A lack of trust amongst project team members
- A fear of conflict directly linked to a lack of passion for the project's key objectives
- An inability to make commitments to anyone associated with the project
- A lack of accountability regarding people's behaviors and the project's deliverables

- An inattention to project results, or expressed in a different way, putting one's individual goals ahead of the project's key objectives

Turning things around becomes a rather tall order for the project manager of a dysfunctional team. By first understanding what a high performance team is all about, the project manager can begin to formulate a detailed plan of action for reversing his project team's poor behavior and results. The essential elements of a high performance team are:

- Having a shared, clear, consistent, and common vision

- Being time oriented, aware, or sensitive

- Communicating very well with each other

- Operating in the "concern" zone which is somewhere between feeling initially comfortable and being very anxious about performing all of the project's lifecycle work activities

- Proactively reviewing the project's quality results

- Involving everyone associated with the project endeavor

- Acting as a self directed or "empowered" group of people when it comes to project matters

- Celebrating their success at key project milestone points and at the project's final conclusion

The project manager needs to take the following approaches to foster his project team into becoming and then operating as a high performance team which are as follows:

- Establish a sense of both urgency along with a very clear sense of direction; this provides the source of energy for developing a passion amongst all of the project team members that is most necessary in order to deliver the project successfully.

- Select members for the project team based upon both their skill and potential, not the force of their personality, nor due to any organizational and political considerations; this will not be easy, nor quick to accomplish well if it is done right.

- Pay very close attention to the early project team member meetings, their decision making process, and any actions that are undertaken. If these do not accomplish what is necessary for success; it is up to the project manager to get directly involved in correcting this situation as soon as possible.

- Set very clear rules of expected behavior.

- Set and seize upon a few immediate performance-oriented tasks and objectives, then build upon this over time.

- Challenge the project team members with fresh facts and information.

- Spend lots of time working and socializing together.

- Exploit the power of positive feedback, recognition, and rewards.

According to Wayne Strider, there are three crucial understandings that one needs to possess in order to be an effective project manager:

- Being well aware of one's self worth

- Being well attuned to other people by acting respectfully and in a careful manner with them at all times

- Being well aware of context — this includes purpose, place, time, roles, conditions, and objects that "self" and "others" operate with. Certain of these aspects of context are self imposed while others are imposed by other people. Keeping one's response within context avoids confusion and acts to reduce any conflict.

The third topic has to do with understanding the concept of emotional intelligence as defined by its creator, Daniel Goleman. people who study leadership indicate that there are three dimensions of intelligence that are applicable to this subject – IQ (Intelligence Quotient), EQ (Emotional Quotient), and MQ (Management Quotient). IQ pertains specifically to critical analysis and judgment, vision and imagination, and possessing a strategic perspective. These characteristics are also known as hard skills and of these three dimensions, IQ has proved to be the least in relative importance in the practice of leadership. EQ encompasses self awareness, emotional resilience, intuitiveness, sensitivity, influencing, motivation, and conscientiousness. These traits are most definitely known as soft skills which will be elaborated upon further later on in this chapter. In terms of the three dimensions, EQ has proved to be second in importance in the practice of leadership and is quite similar to Strider's concept of "self" previously mentioned. MQ includes managing resources (i.e. people, etc.), engaging communication, empowering, developing, and achieving. Just like EQ, these characteristics are soft skills and in terms of relative importance in the practice of leadership rank highest in significance since they help to more frequently contribute to delivering successful project outcomes.

Further, Daniel Goleman developed a model for measuring the Emotional Intelligence of people in terms of four critical skills which are as follows:

- **Self awareness** – understanding one's self and one's emotions (this is rather similar to Wayne Strider's concept of "self" already mentioned above)

- **Self management** – possessing the ability to control one's emotions

- **Social awareness** – recognizing the emotions of others around us (which is rather similar to Wayne Strider's concept of "other" previously mentioned above), expressed as the following:

  - **Empathy** – an ability to understand and relate to the emotions of others

  - **Organizational awareness** – defined by this quote from Daniel Goleman,

Richard Boyatzis, and Annie McKie – "A leader with a keen social awareness can be politically astute, able to detect crucial social networks, and read key power relationships. Such leaders can understand the political forces at work in an organization, as well as the guiding values and unspoken rules that operate among people there."

- **Seeing others clearly** – this is more difficult to perceive since people operating in a work environment too often tend to hide their emotions because they believe this is what is expected of them

- **Emotional boundaries** – where one person's emotions leave off and another's begin

- **Relationship management** – building strong relationships with all project stakeholders, team members, and the project sponsor in order to get important things done very well

By using the proper tools for performing an Emotional Intelligence self assessment, a project manager can begin to determine his own shortcomings, and thus his areas for any and all necessary learning, personal growth, and ultimately his own self improvement. This information is not just useful for the current endeavor that a project manager is working in, but for all future project events. Since projects are unique and temporary by their very nature, it is likely that the members of the project team will be quite different as one goes from project to project. What may have worked well before may not work again because the project circumstances and people involved are very different. A project manager must work very hard at developing strong relationships early in the project lifecycle. Creating and maintaining a positive project team working environment is vital for success in managing people issues over the entire duration of the effort. A project manager needs to develop a sixth sense for anticipating people challengers. Ruth Sizemore House calls this "project radar." She recognizes certain dysfunctional project rules that represent the early warning signs of potential project failure:

- **Groupthink** – striving too hard for consensus. This amounts to forcing people to think, and thus behave, in a certain manner, which is never a desirable thing to do.

- **Burnout** – the depletion of physical and mental (and I would include emotional) resources necessary to attain the objectives of the project. This too often is the result of too many challenging project work activities. The compound effects of stress take their toll at the least opportune time in the project's lifecycle. The main symptoms of burnout are isolation, withdrawal, and exhaustion.

- **Demolition** –the result of escalating infighting. As infighting is allowed to fester and grow, it eventually takes its toll on the project and everyone associated with it. The typical symptoms of demolition are blaming within the project team, sarcasm, an absence of any kind of response, temper explosions, and distortions of information.

• **Collapse** – totally out of bounds project behavior. When project behavior becomes too outrageous, it results in the total collapse of the project. The main symptoms of collapse are false starts and abandonment.

The fourth topic to be covered here has to do with understanding a concept known as Working Styles, facilitated by Dr. Douglas Whittle. In order for a project manager to better understand and effectively deal with all of the people on his project team, recognizing that there are four Working Styles present can be most helpful. Dr. Whittle expresses each of these four Working Styles, developed by Insights®, by using a separate primary color to represent each one – Red, Yellow, Blue, and Green. He says that Red represents individuals who are strong willed, competitive, demanding, determined, and purposeful. But by the same token, such behaviors can become overbearing, aggressive, controlling, intolerant, and driving which can quite clearly put them into conflict with other project team members. Dr. Whittle says that Yellow represents people who are sociable, dynamic, persuasive, enthusiastic, and demonstrative. But under the stress of project work activities, such people can become hasty, frantic, excitable, indiscreet, and flamboyant which can create sources of conflict with others. He says that Blue represents individuals who are questioning, deliberate, cautious, precise, and formal. But by the same token, such people can become suspicious, indecisive, reserved, stuffy, and cold which can produce sources of conflict with other project team members. Finally he says that Green represents people who are caring, patient, sharing, relaxed, and encouraging. But under the stress of project work activities, such people can become bland, docile, reliant, stubborn, and plodding which can cause them to enter into conflict with others.

Another view of Working Styles comes from Robert and Dorothy Grover Bolton. They have identified two dimensions of behavior – Assertiveness and Responsiveness. From this basic concept, they developed a styles grid to be used for one's self assessment – Analytical, Driver, Amiable, and Expressive. According to them, Analytical people tend to be both less assertive and less responsive. By contrast, Drivers tend to be more assertive and less responsive. Amiable individuals tend to be less assertive and more responsive. By contrast, Expressive people tend to be more assertive and more responsive. Both Dr. Whittle and the Bolton's indicate that, while we have a dominant style, none of us operates from just one style all of the time.

The fifth topic pertains to understanding personality types that have been best defined by Myers-Briggs in their well known and established four dichotomies: Introvert and Extravert, Sensing and Intuition, Thinking and Feeling, and Judging and Perceiving. When these are developed further into a matrix, the four dichotomies explode into sixteen distinct personality type indicators, with none of these being best. The chart shown in Table Three, below, displays all sixteen indicators using their abbreviations as now follows:

ISTJ	ISFJ	INFJ	INTJ
ISTP	ISFP	INFP	INTP
ESTP	ESFP	ENFP	ENTP
ESTJ	ESFJ	ENFJ	ENTJ

Table Three – Myers-Briggs Personality Type Matrix

Now as one might expect, an extravert is outgoing and enjoys dealing with people; while an introvert is quiet, reflective, and inner directed. The best way to communicate with an extravert is to get together with them in thinking things through out loud. The best way to motivate an extravert is to have this individual focus upon the relationship aspects of the project at hand. The best way to communicate with an Introvert is to help draw them out and by giving them time to reflect upon the message that is being conveyed. The best way to motivate an introvert is to offer them work that requires extended periods of concentration, even possibly with opportunities to work alone.

A sensing person is pragmatic, practical, and down-to-earth; while an intuitive person is a conceptual, big picture kind of individual. The best way to communicate with a sensing person is to present tangible facts in order to make a point. The best way to motivate a sensing person is to provide them work that has a distinct completion point and can be measured in very concrete terms. The best way to communicate with an intuitive person is to offer an overview presenting concepts that are crucial for discussion. The best way to motivate an intuitive individual is to put them to work on the strategic and design portions of the project.

A thinking person is logical and analytical; while a feeling person is people oriented. The best way to communicate with a thinking person is to present arguments that appeal to a rational analysis of the facts. The best way to motivate a thinking individual is to present this person with tasks requiring quantitative skills, in-depth analysis, or research. The best way to communicate with a feeling person is to talk more from the heart using statements that address values and gut level decision making. The best way to motivate a feeling individual is to allow them to be in roles involving nurturing, supporting, and customer relationship management.

A judging person is orderly, structured, and timely; while a perceiving person is flexible and spontaneous. The best way to communicate with a judging person is to be orderly in presenting your message while keeping the discussion moving toward resolution and closure. The best way to motivate a fudging individual is to permit them to create schedules, budgets, and project closure systems. The best way to communicate with a perceiving person is to allow for open-ended discussion, staying flexible about the agenda. The best way to motivate a perceiving individual is to direct them toward situations requiring trouble shooting.

I have participated in numerous project situations where the Myers-Briggs test was administered at the very start of a project in order to determine the overall nature of the team, as well as to reveal each individual's personal results. We usually discovered a particular type that was dominant in the project team as a whole. This finding was often attributed by the experts to the very nature of the kinds of people who typically gravitate to projects to begin with. Given that projects require a predominance of technical experts this result should really not be so surprising. Often a dominant personality type would also emerge within a sub-group (e.g. financial project team members versus operational or sales project team members). In addition, in a 1996 research paper produced by Max Wideman, entitled "Dominant Personality Characteristics Suited to Running a Successful Project (And What Type are You?)" four types of project leader personality characteristics were derived and placed along two axes. On one axis, labeled Focus, the project leader reflects upon the choice of Problem or Analyze, versus people or Socialize. The problem focused project leader is thoughtful, knowledgeable, and analytical, while the people-focused project leader is frank, demonstrative, and outspoken. On the other axis, labeled Approach, the project leader reflects upon the choice of Receptive or Collaborative, versus Directive. The Receptive project leader's approach is to ask, seek, and obtain consent; while the Directive project leader's approach is to tell, instruct, and dictate. From these two axes were derived four consequent project manager styles: Explorer, Driver, Coordinator, and Administrator.

The Explorer is a person with a vision of the future and projects are the stepping stones to achieving this vision. These people are bold, courageous, and imaginative. The Driver is one who is distinctly action-oriented and hard driving. These individuals are pragmatic, realistic, resourceful, and resolute. They focus on the project's mission and precise project goals. The Coordinator is a person who generally takes a more independent and detached view of their surroundings when the project phase or situation calls for "facilitation." The Administrator is an individual who recognizes the need for stability, typically maximizing repetition on a project in order to optimize productivity. Common characteristics of all four types of project leaders are being "credible, confident, committed, energetic, hard working, and self starting."

This information becomes rather valuable when it is proactively applied to avoiding and resolving people-issue conflicts. The first thing to understand when it comes to conflict is that conflict has both positive and negative aspects to it. The positive aspects are to productively challenge existing beliefs and long standing business practices, to reduce the risk of "groupthink", and possibly to create an opportunity to forge more effective project team relationships, revitalizing team energy, and helping the team to bond. Also, many projects seek substantial changes to business processes which is where the greatest degree of potential ROI exists. When the negative aspects of conflict are not addressed in a productive manner, they can de-motivate project team members and increase interpersonal withdrawal, decrease interpersonal communication, increase cynicism, and adversely affect initiative and the willingness of people to take risks. In addition to any positive and negative aspects of conflict, a project manager needs to consider the project lifecycle phase. The following summary explains this further:

- **Project Initiating phase** – Issues tend to be focused on project priorities, administrative procedures, initial scheduling, and handling the stress of working in a team environment with so many new and different people involved.

- **Project Planning phase** – Issues tend to be focused on project priorities, developing schedules and procedures, working with the functional managers, and handling any disputes that may arise.

- **Project Executing phase** – Issues tend to be focused on achieving the schedule, handling technical challenges, and managing staff resources.

- **Project Closing phase** – Issues tend to be focused on achieving the schedule, the clash of personality styles due to job stress and fatigue, and helping staff deal with the uncertainty related to future work assignments.

Also, in terms of conflict resolution, the project manager should consider if there is any possible lack of information, assess whether any functional issues are present, and determine whether any personality issues exist. There are at least three conflict resolution styles typically used by project managers: avoidance, combativeness, and collaboration. These are to be applied based upon the personal style of the project manager and the particular project phase in which the circumstances surrounding the conflict arise. Finally, one must realize that the practice of project management quite often equates to a great deal of personal stress. This is a rather normal condition; however, critical incidents can happen during a project that may well spike one's stress level for a period of time. The need for work-life balance and the ability to manage one's own personal expectations are absolutely necessary in order to both survive and flourish in such an environment.

In a survey conducted by PMI, a question was posed to their membership: In which process group does the project manager experience the highest degree of stress? The results of this survey are now shown below:

- Project Initiating = 17%

- Project Planning = 33% (the highest result)

- Project Executing = 31% (the second highest result)

- Project Monitoring and Controlling = 18%

- Project Closing = 1% (the lowest result)

Certain important aspects of people working in a project team atmosphere need to be explored a bit further. The three main factors of this form of work life are as follows:

- The degree of subject matter expertise on the project team

- The very nature of the project work itself

- The organizational environment or climate

Of these three factors, the degree of subject matter expertise has proved to be most critical to project team performance. Further, there are four types of project teams as follows:

- **The work group** – Its typical characteristics are top down communication, little work-related communication amongst team members, and the project manager makes all the decisions (a hierarchical approach).

- **The developing team** – Its typical characteristics are that some work-related communication is conducted amongst team members and the project manager makes decisions with input from project team members (an advisory approach).

- **The participative team** – Its typical characteristics are that the project manager and all of the project team members have equal decision-making authority and are constantly in communication with each other (a committee approach).

- **The autonomous team** – Its typical characteristics are that all of the project team members are totally interdependent, have a high level of motivation, and enjoy working with one another.

Understanding the seven most difficult project team personalities is essential to note here as well. These seven are known as the following:

1. **The winner-take-all team** – These teams are in competition with all other project

teams that are currently in existence within the organization, and thus they have a real need to show, at all times, that they are the best at the expense of all others.

2. **The tangent team** – This team has members who establish a project team agenda but they never stick to it and quite often go off on personal agenda tangents.

3. **The social team** – This is a team with members who a track record of working very well with each other, but unfortunately, not of accomplishing their required project goals.

4. **The comedy-show team** – These teams focus on having lots of fun, while the project work is viewed as meaningless, and unimportant.

5. **The whining team** – The team has members who are constantly critical of each other and the work at hand.

6. **The anti-establishment team** – These teams create their own work and project goals.

7. **The self-destructive team** – This team has members who just want the project and the team experience to be over as soon as possible.

Project team conflict also has seven causes which are as follows:

1. Intrapersonal or conflicts within oneself

2. Interpersonal or conflicts between two or more people, or among project teams or other groups of people

3. Structural or conflicts innate to the organizational structure of the work involved

4. Values/beliefs or differences attached to deep-seated emotions

5. Personality or differences in style or behavior

6. Perceptions or differences in view or perspective of the situation, or the specific issue at hand

7. Work methods or disagreements about how to solve problems

By recognizing that different Working Styles and Personality Types really do exist, and that people possess these fundamental human characteristics, a project manager may have a better shot in preparing for, and then in dealing with, people issues whenever they arise. By considering the Working Style and Personality Type of each team member, a greater awareness of how to cope with each of them when resolving a people issue, is more likely to benefit everyone. It may be worth the cost and effort to get project team members to have their Working Styles and Personality Types identified at the very start of a new project. In the long run, this information may help to resolve conflicts with long-lasting results. Understanding the various personal characteristics, the aspects of conflict itself, and how to resolve conflicts will play an important role in dealing with the stress of managing people during a project. By recognizing that these real human behavioral differences exist, a project manager can apply this knowledge to better react to a difficult project situation instead of being completely broadsided by it.

The last topic to be covered here pertains to "soft" skills, which is another critical element that a project manager needs to possess in his personal project management toolkit. Soft skills are those abilities which are demonstrated when:

- Communicating and making presentations of various sorts

- Managing relationships and conflict

- Negotiating and influencing people

- Building teams

- Leading people, to include meetings of various sorts

From this rather short list, we can easily see that soft skills are personal attributes that are directly involved in the practice of project management. In contrast with "hard" skills which tend to be very specific to a certain task or a set of tasks, soft skills are much more broadly applicable. Soft skills can be taught and learned just like other abilities one needs to acquire in life. In terms of project management work activities, a project manager is required to use his soft skills to a far greater extent than his hard skills. A project manager must have excellent listening skills. He or she spends a significant amount of time in meetings, both formal and informal, with people associated in some way with the project. Some of these people are company executives, functional managers, stakeholders, outside vendors and consultants, and fellow project team members. In today's global business environment, some of these people may be from other cultures, generations, races, or sexes, and they may possess critical technical skills needed for the project at hand. Listening then becomes both a challenge, and an absolute necessity. Being able to effectively communicate with these types of people is another crucial soft skill that is used daily by project managers when they make presentations, lead meetings, work to resolve conflicts, and create a great variety of detailed project documents. Remember that communication is a two way process. Just sending information is not nearly enough; it must be fully understood by all of the various receivers. If a project manager fails to do this well, it can become a source of people issues leading to conflict. Further, delivering a clear and consistent message is especially critical to avoiding conflict. If people receive inconsistent project messages, they become rather annoyed and frustrated. It has been said that in the Information Age, we have created many more ways of communicating with each other than ever before in the entire history of mankind; yet, when it comes to performing communications, we are no better than before all of this new technology came along. A study on email communications conducted by Harvard professor Shoshana Zuboff clearly demonstrated that people are much more hostile with each other than if they instead communicated by telephone or, even better yet, face to face.

Managing people is another important soft skill in the practice of project management. The project team must be managed well so that members are focused on accomplishing all of the required work activities in the project plan. Taken together, these critical soft skills must be applied by a project manager in the daily performance of project duties and responsibilities. Periodically, acquiring additional soft skills education and training is just as important as enhancing one's hard skills. PMI requires certified Project Management Professionals (PMPs) to obtain a certain amount of Professional Development Units (PDUs) in order to continue holding on to their certification. They do not dictate what portion of this training must be taken as soft skills versus hard skills, but soft skills training is certainly just as important. Given how important soft skills are in determining the success of projects, one should definitely avail oneself of this training frequently. Even the very best among us needs a refresher from time to time. We must all continuously work to improve our soft skills just as much as we focus on our hard skills.

My final point is that there are two approaches to recognizing and applying soft skills as a project manager. The first is to become a project manager on the basis of one's hard skills, by working one's way up and being in the right place at the right time. That is how many of us became project managers in the first place. When we were lucky, things worked out; however, if we were not, more than likely some people issues caught up with us that exceeded our capacity to deal with them and our project ended up being less successful. You may be surprised by this fact, or you may feel rather overwhelmed by it;

either way, you and your project suffered because of it. I was fortunate enough to fail in my first project before realizing that my project management skills were not nearly developed enough. I needed better soft skills in order to succeed in this profession. A better alternative is to seek out soft skills training prior to becoming a project manager. This definitely would have enabled me to be more successful in my first assignment.

Remember that no one, and no specific tool or set of tools, can guarantee a successful project outcome every time and in every project situation. There are too many variables and circumstances for that to realistically happen. However, this book is all about understanding practical ways to increase the success rate of projects, and soft skills training can do that very thing. Good project managers are totally committed to continuous learning, self improvement, and acquiring relevant feedback. Even if they do not receive any PDU credits, project managers would still seek out additional soft skills training. By using one's creativity and by being flexible in resolving project problems, a project manager can best demonstrate soft skills for the benefit of all parties, and for the project itself.

CHAPTER EIGHT

Epilogue

Back in Part One of this book, I used two key statistics to portray three project plays demonstrating just how much people challenges can lead to inconsistent and negative project outcomes. In the section on lessons learned following these three plays, we see that all three project situations had a great many things in common. It is my contention that if people are the greatest cause of project challenge, then just maybe they might also be part of the solution. Sometimes fighting fire with fire is simply the best solution.

People generally seek to do a good job. Projects are the way that most work is performed in business today. It seems appropriate to believe that good people will want to be part of such project work, and that they will desire to perform it by doing their very best. To do so requires a three-legged stool supported by technical competency, a proper understanding of project management methods and tools, and good interpersonal (i.e. soft) skills. My position in this book is that all three matter a great deal in order to arrive at a successful project outcome. However, it is my further contention that people or soft skills have been neglected too long in project manager training. Leading people through the lifecycle of a project is like no other form of work. Project leadership has to do with leading a diverse group of technical experts, over whom the project manager has no formal authority. Add into this mix the uncertainty of learning to use new technologies, a tight schedule and budget, global business operations, and a demanding customer, and we have a recipe for project complexity, and the resulting people challenges.

The good news in all of this is that just like mastering a technical competency or a project management methodology, soft skills and project leadership can be learned, which may result in greatly improved project outcomes. We give out university degrees in technical competencies and increasingly we hand out certifications in project management as well. The development of strong people skills should not be lacking. As the global field of project management continues to grow, this third leg of the stool must be addressed proportionately to compensate for the high rate of information technology project failure. Dr. Al Zeitoun has stated that "The artistic side of the project manager as a leader is what allows project managers to be who they could best be. Projects do not succeed in creating the 'wow' for customers and society only because the best technology was used; they succeed because the collaboration of minds towards a well defined objective took place as directed through proper leadership. It is that side of the project manager that makes the miracle happen".

The Project Management Institute's *PMBOK Guide (A Guide to the Project Management Body of Knowledge)* does a tremendous job at providing project managers and project team members with great insight into all of the processes, and technical aspects of this growing global profession known as project management. It does not, however, provide practical information about dealing with people issues and the handling of such conflicts. The *PMBOK Guide* speaks to this subject matter only at a very superficial level. It is my hope that this book will help my fellow project management professionals to understand better the importance of people skills in the achievement of project leadership, and to achieve more successful project outcomes. In so doing, the success rate for projects will definitely see an increase well above 50% in due course. It is imperative for all parties involved with project activities to see that this matter is addressed without any delay. Time is of the essence given that projects are growing exponentially in terms of their numbers and their degree of complexity. As I said earlier, project work is becoming the norm in today's working world. This is primarily due to the fact that a greater proportion of our economy is now part of the global economy, with all of the many challenging business issues that this all entails. Further, increases in the application of new technologies are occurring very rapidly and will only continue to do so. The interactions between introducing and using these new technologies effectively within an existing business environment clearly present significant people and technical challenges. When you factor in that 80% of all project issues are due to people, however, it is up to people to see that this result is changed for the better. This includes everyone: project managers, project sponsors, stakeholders, and all project team participants. Project charters, plans, methods, and technical tools will not by themselves get a project accomplished – only people, working well together, can do so. The best advice given about buying real estate by its professionals is "Location, location, and location". In the project management profession, the mantra ought to be "people, people, and people".

In summary, I now leave you with a brief personal story I wish to share. Some ten years ago, when I was teaching an Introduction to Project Management university course, a student asked me during the last class a very interesting question. It directly related to my many comments in our previous fourteen weekly class sessions about the inherent stress involved with performing project management activities. This student's question was simply, "If the responsibility for performing the job of a project manager is so great, why anyone would actually choose to do it?" My answer at the time was that certain people, such as me, get turned on by the many challenges that projects generate, and they proactively seek to participate in projects and then work very hard to accomplish them successfully. I still believe in what I said back then but I would now add that understanding the significance that people issues have on projects requires that a project manager learns and masters additional soft skills that go way beyond purely process and technical matters in order to achieve real and lasting success in both the project's outcome and one's career. I have watched dozens of other project team leaders and project managers over the years in which I have been practicing the profession of project management. Some of these people have lacked formal education and certification in project management, while others were on par or even somewhat better than me in this regard. In spite of this, some of these people were successful in their project endeavors while others were not. Two key differentiators that stand out were their degree of desire (i.e. passion) for the endeavor and their ability to adapt to the project environment they were given, especially in regards to all of the people involved in the effort. Above all else, it is the ability to deal effectively with the people associated with a project that makes the job of project management so easy or so very hard, as well as such fun or so very stressful. Accomplishing great new things is what makes the job of being a project manager so exciting and personally rewarding, regardless of the actual compensation that is involved. Performing this important job with the help and support of other people is something every project manager strives for in their daily practice.

By making soft skills learning a priority, project managers will become much more effective and successful. The real secret of project leadership is learning how to build and grow one's interpersonal relationships.

In closing, I personally believe that the most critical project leadership ability, bar none, is to possess and regularly display to all a good attitude balanced with an appropriate sense of humor. Without this, one is going to have a much more difficult project journey. No matter what is thrown in your path, if your attitude demonstrates that you can take things in stride and move forward, people will be more willing to work with you in achieving the desired project results, no matter what the various challenges might be – people, process, or technological – but especially people issues, above all else. It never ceases to amaze me just how great people challenges can be and how they never stop coming each and every day that I practice the profession I have come to love: project management.

The Final Word

"There is nothing more difficult to take in hand,
more perilous to conduct, or more uncertain in its
success, than to take the lead in the introduction of
a new order of things. Because the innovator has
for enemies all those who have done well under the
old conditions, and lukewarm defenders in those
who may do well under the new."

A quote from Nicolo Machievelli's
book *The Prince*, dated 1515.

References

1. All Project Management quotations came from www.Wikiquote.org

2. Project Management Institute (PMI) Strategic Plan – per Kathleen Romero's presentation at the December 12, 2006 PMI Birmingham, Alabama chapter dinner meeting

3. O'Neill Project Management research study of 1999 – where it was determined that project managers spend 70% of their time on non-value added project activities – "Essential People Skills for Project Managers", page 154

4. The Standish Group – CHAOS research studies of 1994 and 2004

5. Project Management Body of Knowledge (PMBOK), 3rd edition – Project Management Institute (PMI) – the five stages of the project lifecycle, page 38

6. PM Network magazine of August 2006 – "Dysfunction Junction" by Ann C. Logue, pages 78-81

7. Dr. Qian Shi and Dr. Jianguo Chen – "The Human Side of Project Management: Leadership Skills", page 4

8. Dr. Shoshana Zuboff – "In the Age of the Smart Machine", chapter 10

9. Organizational Culture – from www.1000advices. com/guru by Vadim Kotelnikov

10. "Internal Politics at Work Place" – from http:// EzineArticles.com/?expert=Arvind_Katoch

11. Dr. Douglas Whittle – "Using Working Styles to Help Your Team Cross the Finish Line" – presentation made at the ERP WIS conference in Orlando, FL in October 2006

12. Robert Bolton and Dorothy Grover Bolton – "People Styles at Work", chapter 4, pages 16-23

13. Myers-Briggs Type Indicator (MBTI) – from www.Wikipedia.com

14. Max Widemen – "Dominant Personality Traits Suited to Running Projects Successfully (And What Type are You?)" – from www.maxwidemen. com/papers/personality/review1980.htm

15. Three conflict resolution styles by Robert K. Wysocki – "Effective Project Management", 3rd edition, pages 196-197

16. Four models of project teams – "Choose the Right Team Model" by Gary Topchik, June 14, 2007 from www.projectsatwork.com/articles/ articlesPrint.cfm?ID=236920

17. Seven most difficult project team personalities and causes of conflict by Gary Topchik – "The

First-Time Manager's Guide to Team Building", AMACOM, 2007

18. Dr. Al Zeitoun – from http://www.maxwidemen. com/guests/balanced_blend/done.htm

19. Dysfunctional teams – "The Turnaround Artist" by Sarah Fisher Gale – PM Network magazine of October 2007, pages 25-30

20. High performance teams – from http:// highperformanceteams.org/combined.htm

21. Context – "Powerful Project Leadership" by Wayne Strider – Introduction, pages 7-9

22. "Emotional Intelligence for Project Managers" by Anthony Mersino, chapter 5

23. Leadership from www.Wikipedia.com

24. Management versus Leadership from www. Wikipedia.com

25. Definition of a Project – PMBOK, pages 5-6

26. Leadership Styles – RMC's (Rita Mulcahy's) PMP Exam Preparation book, 5th edition, page 283

Bibliography

Project Management

House, Ruth Sizemore. *The Human Side of Project Management.* Addison-Wesley Publishing Company, Inc., 1988.

Keogh, Jim. *Project Planning and Implementation.* Pearson Custom Publishing, Custom, 1994.

Shi, Dr. Quian. Chen, Dr. Jiamugo. *The Human Side of Project Management: Leadership Skills.* Newtown Square, PA: Project Management Institute, Inc., 2006.

Whitten, Neal. *Neal Whitten's No-Nonsense Advice for Successful Projects.* Management Concepts. 2005.

Wysocki, Robert K. *Effective Project Management.* 3rd edition. Wiley Publishing Inc., 2003.

Leadership

Flannes, Dr. Steven W. and Levin, Dr. Ginger. *Essential People Skills for Project Managers.* Management Concepts, 2005.

Johnson, Jim. *My Life is Failure.* The Standish Group International, Inc., 2006.

Kloppenborg, Timothy J., Shriberg, Arthur; and Vankataman, Jayashreeu. *Project Leadership.* Management Concepts, 2003.

Lapid-Bogda, Dr. Ginger. *What Type of Leader are You?* McGraw-Hill, 2007.

Mersino, Anthony. *Emotional Intelligence for Project Managers.* American Management Association, 2007.

Strider, Wayne. *Powerful Project Leadership.* Management Concepts, 2002.

Soft Skills

Geoghegan, Linda and Dulewicz, Victor. "Do Project Managers' Leadership Competencies Contribute to Project Success?" *Project Management Journal.* December 2008.

Project Teams

Bolton, Robert and Bolton, Dorothy Grover. *People Styles at Work.* American Management Association, 1996.

Parker, Glen M. *Cross-Functional Teams.* Jossey-Bass Publishers, Inc., 1994.

Thompson, Leigh; Aranda, Eileen; and Robbins, Stephen P. *Tools for Teams.* Pearson Custom Publishing, 2000.

Some Project Management Quotations

"Coming together is the beginning. Keeping together is progress. Working together is success."
(Henry Ford)

"If there were no problem people there'd be no need for people who solve problems."

"A project is one small step for the project sponsor, one giant leap for the project manager."

"Too few people on a project can't solve the problems – too many create more problems than they solve."

"A project gets a year late one day at a time."

"Powerful project managers don't solve problems, they get rid of them."

*"The first **90%** of a project takes **90%** of the time while the last **10%** takes the other **90%**."*

"If you fail to plan, you are planning to fail."

"Activity is not achievement."

"All project managers face problems on Monday mornings – good project managers are working on next Monday's problems."

"Any project can be estimated accurately (once it's completed)."

"Everyone asks for a strong project manager – when they get him they don't want him."

"It's often said that hard skills will get one an interview, but one needs soft skills to first obtain and then keep the job."

"I can't imagine a person becoming a success who doesn't give his game of life everything he's got."

"You cannot make up for 'soft' skills with hard work."

"The most successful project managers have perfected the skill of being comfortable being uncomfortable."

About the Author

 Richard Bernheim has been a senior management consultant in three of the top four consulting firms (Ernst & Young, Deloitte & Touche, and Price Waterhouse) where he has specialized in implementing financial and cost management information systems such as SAP R/3. He was featured in the 1987 (25th) edition of *Who's Who in the Finance Industry in America.*

A popular educator and speaker, he has been a part-time professor of financial management and project management at Montgomery County Community College, Temple University, and University of Phoenix. He has also been a featured speaker at various conferences, at PMI chapter meetings, and at other professional organizations since 1983.

An accomplished writer, Bernheim has authored 10 published manuscripts for professional organizations and has proofread and revised several college-level financial and cost accounting textbooks.

He holds a BA in political science, an MBA in finance, a PMP from the Project Management Institute, and is certified in various SAP R/3 implementation methodologies and ASAP Project Management.

To obtain additional information, including related materials, please check out the author's Website at the following Internet address:

www.BeaSmartPM.com

Did you like this book?

If you enjoyed this book, you will find more interesting books at

www.MMPubs.com

Please take the time to let us know how you liked this book. Even short reviews of 2-3 sentences can be helpful and may be used in our marketing materials. If you take the time to post a review for this book on Amazon.com, let us know when the review is posted and you will receive a free audiobook or ebook from our catalog. Simply email the link to the review once it is live on Amazon.com, with your name, and your mailing address—send the email to orders@mmpubs. com with the subject line "Book Review Posted on Amazon."

If you have questions about this book, our customer loyalty program, or our review rewards program, please contact us at info@mmpubs.com.

Managing Smaller Projects: A Practical Approach

So called "small projects" can have potentially alarming consequences if they go wrong, but their control is often left to chance. The solution is to adapt tried and tested project management techniques.

This book provides a low overhead, highly practical way of looking after small projects. It covers all the essential skills: from project start-up, to managing risk, quality and change, through to controlling the project with a simple control system. It cuts through the jargon of project management and provides a framework that is as useful to those lacking formal training, as it is to those who are skilled project managers and want to control smaller projects without the burden of bureaucracy.

Read this best-selling book from the U.K., now making its North American debut. *IEE Engineering Management* praises the book, noting that "Simply put, this book is about helping people control smaller projects in a logical and effective way, and making the process run smoothly, and is indeed a success in achieving that goal."

Available in paperback format. Order from your local bookseller or directly from the publisher at

http://www.mmpubs.com/msp

Winston Churchill: The Agile Project Manager

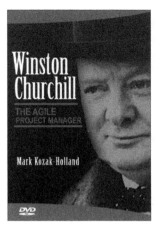

Today's pace of change has reached unprecedented levels only seen in times of war. As a result, project management has changed accordingly with the pressure to deliver and make things count quickly. This recording looks back at a period of incredible change and mines lessons for Project Managers today.

In May 1940, the United Kingdom (UK) was facing a dire situation, an imminent invasion. As the evacuation of Dunkirk unfolded, the scale of the disaster became apparent. The army abandoned 90% of its equipment, the RAF fighter losses were deplorable, and over 200 ships were lost.

Winston Churchill, one of the greatest leaders of the 20th century, was swept into power. With depleted forces and no organized defense, the situation required a near miracle. Churchill had to mobilize quickly and act with agility to assemble a defense. He had to make the right investment choices, deploy resources, and deliver a complete project in a fraction of the time. This recording looks at Churchill as an agile Project Manger, turning a disastrous situation into an unexpected victory.

ISBN: 1-895186-50-1   (Audio CD)
ISBN: 1-897326-38-6   (DVD)

http://www.mmpubs.com

Managing Agile Projects

Are you being asked to manage a project with unclear requirements, high levels of change, or a team using Extreme Programming or other Agile Methods?

If you are a project manager or team leader who is interested in learning the secrets of successfully controlling and delivering agile projects, then this is the book for you.

From learning how agile projects are different from traditional projects, to detailed guidance on a number of agile management techniques and how to introduce them onto your own projects, this book has the insider secrets from some of the industry experts – the visionaries who developed the agile methodologies in the first place.

ISBN: 1-895186-11-0   (paperback)
ISBN: 1-895186-12-9   (PDF ebook)

Also available in ebook formats. Order from your local bookseller, Amazon.com, or directly from the publisher at

http://www.mmpubs.com

Lessons from the Ranch for Today's Business Manager

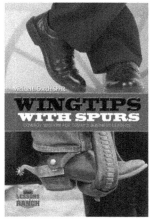

The lure of the open plain, boots, chaps and cowboy hats makes us think of a different and better way of life. The cowboy code of honor is an image that is alive and well in our hearts and minds, and its wisdom is timeless.

Using ranch based stories, author Michael Gooch, a ranch owner, tells us how to apply cowboy wisdom to our everyday management challenges. Serving up straight forward, practical advice, the book deals with issues of dealing with conflict, strategic thinking, ethics, having fun at work, hiring and firing, building strong teams, and knowing when to run from trouble.

A unique (and fun!) approach to management training, Wingtips with Spurs is a must read whether you are new to management or a grizzled veteran.

ISBN: 1-897326-88-2   (paperback)

Also available in ebook formats. Order from your local bookseller, Amazon.com, or directly from the publisher at

http://www.mmpubs.com

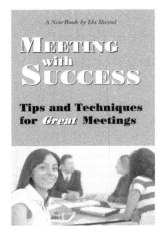

A New Book by Ida Shessel

**MEETING
with
SUCCESS**

**Tips and Techniques
for *Great* Meetings**

Are People Finding Your Meetings Unproductive and Boring?

Turn ordinary discussions into focused, energetic sessions that produce positive results.

If you are a meeting leader or a participant who is looking for ways to get more out of every meeting you lead or attend, then this book is for you. It's filled with practical tips and techniques to help you improve your meetings.

You'll learn to spot the common problems and complaints that spell meeting disaster, how people who are game players can effect your meeting, fool-proof methods to motivate and inspire, and templates that show you how to achieve results. Learn to cope with annoying meeting situations, including problematic participants, and run focused, productive meetings.

ISBN: 1-897326-15-7   (paperback)

Also available in ebook formats. Order from your local bookseller, Amazon.com, or directly from the publisher at

http://www.mmpubs.com/

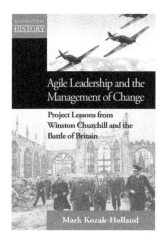

Agile Leadership and the Management of Change: Project Lessons from Winston Churchill and the Battle of Britain

Around the turn of the millennium, there was a British poll that asked who was the most influential person in all of Britain's history. The winner: Winston Churchill. What distinguished him were his leadership qualities: his ability to create and share a powerful vision, his ability to motivate the population in the face of tremendous fear, and his ability to get others to rally behind him and quickly turn his visions into reality. By any measure, Winston Churchill was a powerful leader.

What many don't know, however, was how Churchill used his leadership skills to restructure the British military, government, and even the British manufacturing sector to get ready for an imminent enemy invasion in early 1940.

Learn how Churchill acted as the head project manager of a massive change project that affected the daily lives of millions of people. Learn about his change management and agile management techniques and how they can be applied to today's projects.

ISBN: 9781554890354   (paperback)
ISBN: 9781554890361   (PDF ebook)

http://www.mmpubs.com/

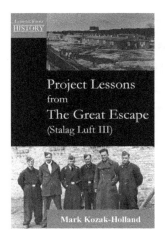

Project Lessons from The Great Escape (Stalag Luft III)

While you might think your project plan is perfect, would you bet your life on it? In World War II, a group of 220 captured airmen did just that – they staked the lives of everyone in the camp on the success of a project to secretly build a series of tunnels out of a prison camp their captors thought was escape proof.

The prisoners formally structured their work as a project, using the project organization techniques of the day. This book analyzes their efforts using modern project management methods and the nine knowledge areas of the *Guide to the Project Management Body of Knowledge* (PMBoK).

Learn from the successes and mistakes of a project where people really put their lives on the line.

ISBN: 1-895186-80-3   (paperback)

Also available in ebook formats.

http://www.mmpubs.com/escape

BILL RICHARDSON

Thinking on Purpose for Project Managers: Outsmarting Evolution

When you're facing down a lion on the open savannah, automatic reactions hardwired into your system through eons of evolution can save your life. However, when you're trying to impress the CEO across a boardroom table, those same responses can cost you big time.

So, how do you overcome your automatic reactions, retrain your brain, and outsmart evolution? By learning and using the techniques revealed in *Thinking on Purpose for Project Managers: Outsmarting Evolution.*

ISBN: 9781554890255 (paperback)

Also available in ebook formats. Order from your local bookseller, Amazon.com, or directly from the publisher at **http://www.mmpubs.com**

Are We There Yet? Diary of a Project Manager

As project management techniques become ubiquitous in the workplace, more and more people find themselves challenged by bureaucracy, unruly team members, and irrational customers. This compelling (and sometimes humorous) book follows the day-to-day trials and tribulations of a team working to commercialize an innovative new product. The team faces commonplace issues such as corporate leadership and strategy; impacts on sales and marketing; finance; operations; design; production and the supply chain. Presented in diary format to make for an engaging read, each chapter ends with a lesson on what could have improved the team's performance, focusing on the four "Ps" of project management: processes, people, parts, and phenomena.

Sneak a peek in this diary to discover the Good, the Bad, and the Utterly Random concerning one of the most essential, unpredictable and unsung responsibilities in business.

ISBN: 9781554890293   (paperback)

Also available in ebook formats. Order from your local bookseller, Amazon.com, or directly from the publisher at **http://www.mmpubs.com**